Fourth Edition, Revised

A Primer of
OILWELL
DRILLING

by Ron Baker

Published by
PETROLEUM EXTENSION SERVICE
The University of Texas at Austin
Austin, Texas

in cooperation with
INTERNATIONAL ASSOCIATION
OF DRILLING CONTRACTORS
Houston Texas

1979

Catalog No. 2.00040
ISBN 0-88698-080-1

CONTENTS

FOREWORD

The drilling of oil and gas wells is a vital part of the petroleum industry, for without a well it is all but impossible to obtain hydrocarbons that reside deep within the earth. Because an understanding of drilling is so important to the comprehension of the entire industry, this edition of *A Primer of Oilwell Drilling* is offered. The *Primer* is just what its name implies — a first reader of the oilwell drilling business.

PETEX published the first edition of the *Primer* in 1951, the second in 1957, and the third in 1970. This fourth edition was first released in 1979 and has been reprinted many times. It still has the purpose that brought the original manual into existence: to give nontechnical persons a simple insight into the world of drilling an oil well and to acquaint them with some of the terminology, procedures, and problems encountered in drilling holes in the ground.

Ron Baker, Director
Petroleum Extension Service

PREFACE

The idea behind this book is to explain in simple words and pictures some of the basic things it takes to drill an oil or gas well. It is not easy to keep a book about drilling simple. A drilling rig has so many parts and there are so many operations and techniques involved that a really thorough treatment would require several volumes and a reader with the understanding of an advanced engineer or scientist.

The language of the drilling industry abounds with special words and phrases; there is a lot of jargon that folks who are new to the business just cannot be expected to understand. Whenever you are explaining anything about drilling, it is really easy to slip into using this special language. After all, jargon is mainly a kind of shorthand for those in the know. Those out of the know cannot figure out what is going on.

So, while we haven't eliminated the jargon entirely (most of it is interesting, colorful, and essential to learn if you're going to talk drilling), we have tried to define each special word the first time it comes up and then incorporate it in the text thereafter.

Now, those who already know drilling are going to find that we've left some things out. How do you cover *everything?* To those who are offended because we neglected to mention a process or piece of equipment that is especially near and dear to them, we offer out sincere apologies. But keep in mind that the intent of this book is to give beginners a basic understanding of drilling without bowling them over with extreme detail. Of course, there's always room for improvement, and suggestions on how this primer can be made better will be gratefully accepted. And if we made a mistake somewhere, we definitely want to know about it.

Putting a book like this primer together was not easy. It took the skills of just about everyone on the PETEX staff. Paula Floyd painted the cover and Mark Donaldson illustrated the text with help from David Pound. Dan Diener and Mel Lange took many of the excellent photos. Of course, no book is completed until it is typed, edited, proofread, pasted up, and in general, made ready for the printer. Deborah Caples, Debbie Warden, Elsa Calderón, Janice Kozlowski, Martha Greenlaw, and Mildred Gerding handled these essential chores in their usual superlative manner.

Warm thanks go to Ed McGhee of the International Association of Drilling Contractors for his constructive comments. And last, but not least, we wish to thank the numerous people from the many service, operating, and drilling companies without whose help and generosity this book could never have come into being.

INTRODUCTION

One Sunday afternoon in the latter part of August 1859, a blacksmith known as Uncle Billy Smith sauntered over to a well he was drilling near Oil Creek just outside of Titusville, Pennsylvania. In those days, well drillers took Sundays off, and Uncle Billy was probably curious to see if anything had happened while he was in church. He'd started the well back in April, working for an unemployed railroad conductor named Edwin L. Drake. Drake, known to most as "Colonel," had hired Uncle Billy to do something that had never been done before in the United States: drill a well into the earth for the express purpose of finding oil. It is not known what thoughts Uncle Billy had when he found the hole he'd helped drill full of oil. What *is* known is that the Drake well heralded the beginning of an era: the petroleum era.

Of course, drilling operations have undergone vast changes since the Drake well (fig. 1). The Drake well bottomed out at 69½ feet, not a particularly monumental depth even for 1859. Indeed, Colonel Drake bragged that he was prepared to drill as deep as 1,000 feet if necessary. In 1974, after 504 days of drilling, Loffland Brothers Rig 32, drilling for Lone Star Producing Company in southwestern Oklahoma, bottomed out a hole at 31,441 feet, a record depth in the United States that still stands today (fig. 2). At present, the Soviet Union is drilling a well on the Kola Peninsula, to the east of Finland, that has passed 37,000 feet in depth. It is scheduled to go to 49,212 feet. (That's over 9 miles, straight down!) Unlike Drake, the Oklahoma drillers did not strike petroleum, nor have the Soviets.

Now let's go back in time once again—this time to Texas. It is January 10, 1901, and a group of drillers—headed up by an Austrian-born mining engineer named Anthony Lucas—have just installed a new drilling bit on the end of their string of drill pipe. They've been at the drill site, a place called Spindletop that lies about 4 miles south of Beaumont, since October 27, 1900.

Figure 1. A restoration of the Drake well at Titusville, Pennsylvania

1

Figure 2. Loffland Brothers Rig 32 drills the deepest U.S. well (31,441 feet) in Southwestern Oklahoma.

They begin lowering the new bit to bottom, and, after getting about 700 feet of drill pipe into the 1,020-foot hole, the well starts spewing drilling mud. The mud, a liquid concoction used to carry rock cuttings out of the hole, drenches the rig floor and shoots up into the derrick.

After a short while, the flow stops, so the crew returns to the rig to clean it up. Suddenly, mud begins to erupt again and is quickly followed by a geyser of oil, which gushes up to 200 feet above the 60-foot high derrick. All the drill pipe is blown out of the hole and falls in the derrick and onto the ground. Nobody is hurt. Instead, they are elated. Here is a well flowing oil at the rate of 84,000 barrels per day! Up to now, 50 barrels per day was considered to be a phenomenal amount, but here is one flowing nearly 1,700 times that amount!

2

The great Lucas well at Spindletop ushered in the petroleum industry as it is known today (fig. 3). Among other things, it established the usefulness of rotary drilling as a viable means of drilling holes into the ground. Up to 1900 (even though several wells had already been

Figure 3. The Lucas well, Spindletop's gusher, proved the value of using a rotary rig to drill wells.

drilled by turning or rotating the drill pipe and bit on bottom to "make hole"), the dominant method of drilling was the cable-tool method (fig. 4). In cable-tool drilling, the bit is not rotated; instead, the bit is attached to wire rope (or cable) and is picked up and dropped, picked up and dropped, over and over until the hole achieves the desired depth. Each time the bit hits bottom, it punches through a little more formation. From time to time, drilling has to stop and the

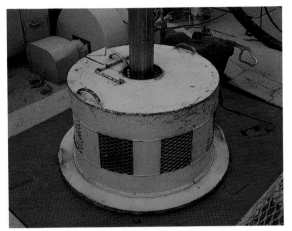

Figure 5. A modern rotary table turns or rotates to make the bit drill. This one has a guard over it.

cuttings made by the bit bailed out of the hole before drilling can be resumed.

On the other hand, rotary drilling features a rotating machine, called the rotary or rotary table, through which pipe (instead of wire rope) is run (fig. 5). The bit is attached or "made up" on the drill pipe. When the rotary is engaged, it rotates the pipe and bit, and the hole is drilled by the turning bit. Also, rotary drilling offers another big advantage over cable-tool drilling: drilling fluid, usually called "mud," can be pumped down through the drill pipe, out the bit, and back up to the surface. As a result, the rock cuttings made by the bit can be lifted by the mud and carried up to the surface for disposal. No bailing of cuttings is necessary, so drilling does not have to stop as often. Circulating drilling fluid also has many other advantages.

In any case, Spindletop was a landmark event in the petroleum industry, both in terms of the prodigious quantities of oil it discovered and in terms of establishing the usefulness of rotary drilling. Today, almost all drilling is done by the rotary method even though in certain areas quite a few wells are still being drilled with cable-tool rigs. (Drilling by the cable-tool method is a very efficient way of making hole. Unfortunately, there is no way that drilling fluid can be circulated, and this disadvantage makes cable-tool drilling rigs a vanishing breed.) This book will tell the story of the people and equipment it takes to drill a well with a rotary drilling rig.

Figure 4. The rig floor of an old cable-tool rig at work in Kansas.

PETROLEUM PEOPLE

No treatise on rotary drilling is complete without some discussion about the people who are involved in drilling and producing the most popular energy sources the world has ever known—oil and gas. The oil business has grown from a small number of promoters in Pennsylvania, producing and marketing "rock oil" in competition with whale oil, to the millions who now participate in the finding, drilling, producing, refining, and marketing of oil and gas the world over. And this is not to mention the large number of everyday citizens who own stock in petroleum and petroleum-related companies. In addition, the scientific discoveries and technological developments stimulated by the needs of the petroleum industry furnish the livelihood for thousands of other workers. And, of course, almost every citizen of almost every country is a petroleum consumer.

THE COMPANIES

Within the petroleum industry, the drilling of wells is of paramount importance: with few exceptions, there is no way to obtain oil and gas unless a well—a simple (yet complex) hole in the ground—is drilled. The drilling phase of petroleum, then, becomes a subject worth considering in detail. Basically, the people directly involved in drilling are employed by operating companies, drilling contractors, and various service and supply companies.

Operating companies are the financiers of the industry and are the principal users of the services provided by drilling contractors and service companies. An operating company, often called simply an operator, is a person or

company who actually has the right to drill and produce petroleum that may exist at a particular site. The operator acquires the right by buying or leasing it from the person or persons who own the land under which petroleum may exist. An operator can be a major company, such as Exxon, Shell, Chevron, Texaco, Mobil, ARCO, Phillips, Conoco, or Standard, to name but a few. Or an operator can be an "independent," a relatively small and largely unintegrated company or individual who produces and sells oil and gas but is not engaged in the transporting, refining, or marketing of it. A major oil company, on the other hand, is a large, integrated company that not only produces oil and gas, but transports it from the field to the refinery and plant, refines or processes it, and sells the products to the general public and other consumers.

Both major and independent companies initiate the sequence of events that leads to the drilling of oil and gas wells in the United States. In other countries, the oil industry is often nationalized; that is, a special ministry or department of government controls the country's petroleum affairs. However, major U.S. oil companies are often invited to participate in oil and gas ventures abroad because of their experience, equipment, and expertise.

Operating companies, both large and small, have found it more expedient to utilize the men, equipment, and skills and experience of *drilling contractors* to perform the actual drilling of a well. Thus, about 98 percent of all drilling in the U.S. is done by drilling contractors. Like an operating company, a

Figure 6. Roustabouts prepare a load of supplies on the deck of a boat prior to transferring the load to an offshore rig.

drilling company may be small or quite large. Regardless of size, the main job of the contractor is to drill a hole to the depth and specifications set by the operating company. Usually, the operator sends out an invitation to bid to several drilling contractors who he knows have the equipment and integrity to drill the well to the specifications stated in the bid invitation. Drilling contractors must be competitive in prices and services to stay in business. If a contractor is awarded the contract, it is because his is the best bid. It may not be the lowest bid, since past performances and proven capacity are also factors to be considered.

The contract for making hole goes to a drilling contractor, but the operating company is usually the one that awards other contracts to the *service and supply companies.* (It should be noted, however, that the contractor also relies on supply companies to provide him with tools and services he needs to drill and to keep his rigs in good working condition.) Since it takes various expendable supplies such as drilling mud and bits to drill a well, the operator and contractor buy the materials from companies that specialize in a particular service.

Also, specialists such as exploration geophysicists (people who work to find petroleum deposits) often operate under contracts separate from that made with the contractor. Indeed, after all the various attendant services needed to drill a well are summed up, there could be as many as a hundred service and supply companies working at the drill site at one time or another. The coordination of all these services is supervised by the operating company, its managers, and engineers. Full cooperation and assistance of the drilling contractor is required in almost every instance.

THE DRILLING CREW

Although the number of persons (women as well as men) on a drilling crew varies from rig to rig, most crews working on land rigs consist of a toolpusher, a driller, a derrickman, and two or three rotary helpers (also known as floormen or roughnecks). Sometimes the most experienced rotary helper will be designated as a motorman. And sometimes additional crewmen are hired as rig mechanics and rig electricians. While not an actual member of the rig crew, a representative of the operating company is usually on the site at all times. He is often called the company man, although more formally he is called the company representative.

Offshore, the contractor employs all the people that a land contractor hires, but, because of the more complex nature of the offshore operation, he hires many additional persons as well. For example, a number of *roustabouts* (fig. 6) will work under the direction of the *head roustabout,* who is very often the *crane operator* as well (fig. 7). Roustabouts handle the equipment and supplies that are almost constantly being supplied to the rig from a shore base. The crane opeator has the skills to operate the large cranes that are used on most offshore rigs to load and unload supplies.

Figure 7. A crane operator works in an offshore crane mounted on a platform rig.

Whether on land or offshore, the **company man** is in direct charge of all the company activities on the drilling location. He plans the strategy for the drilling of the well, orders the needed supplies and services, and is responsible for making decisions that affect the progress of the well.

The **toolpusher** is the drilling contractor's top man on location at the drill site. He supervises all drilling operations and coordinates company and contractor affairs (fig. 8). He, like the company man, lives on location in a trailer or portable building and is on call 24 hours a day.

The **driller** is under the direct supervision of the toolpusher and is the overall supervisor of the floormen. He operates the drilling machinery on the rig floor. If any one person can be described as being directly responsible for the actual drilling of the hole, it is the driller. From his position at a control console on the rig floor, he manipulates the levers, switches, brakes, and other related controls that keep the bit on bottom drilling (fig. 9).

Figure 9. The driller, at his position on the rig floor, operates the drilling machinery.

Figure 8. The toolpusher and company man discuss a problem on the rig floor.

The **derrickman** works on the monkeyboard, a small platform located up in the derrick at a level of the upper end of a length of drill pipe (usually about 90 feet). When pipe is taken out of the hole (tripped out), he handles the upper end of the pipe, guiding it to and from the special equipment it takes to run pipe in and out of the hole (fig. 10). When the bit is on bottom drilling and all the pipe is in the hole, the derrickman is responsible for keeping tabs on the drilling fluid and for maintaining or repairing the pumps and other equipment needed to circulate the fluid.

Figure 10. The derrickman handles the upper end of the drill pipe from his position on the monkeyboard.

The *rotary helpers* (at least two, usually three, and sometimes four) are responsible for handling the lower end of the drill pipe when it is being tripped in or out of the hole (fig. 11). Also, they handle the large wrenches, called *tongs*, that are used to screw or unscrew (make up or break out) the individual lengths of pipe. Besides these duties on the rig floor, rotary helpers often maintain equipment, keep it clean and painted, and in general lend a helping hand wherever it may be needed on the location.

Figure 11. Three rotary helpers handle pipe during a trip.

The *motorman* is responsible for keeping the engines that provide the power for the drilling equipment on the rig in good working order. He maintains the engines, adding oil and coolant when necessary (fig. 12). Not all rigs have a motorman. Sometimes the rotary helpers or derrickmen, under the driller's supervision, attend to the maintenance needs of the rig engines.

Figure 12. A motorman lubricates a part on one of the rig engines. Note the ear protection device he wears to protect his hearing from the engine roar.

The *rig mechanic*, if a rig has one, is an all-round handyman when it comes to anything mechanical on the rig. He may make minor repairs on the engines, small pumps, and various other machinery on and around the rig.

The *rig electrician* maintains and repairs the electrical generating and distribution system on the rig. He may make minor repairs to generators or electric motors, inspect and maintain the rig's electrical wiring, and maintain the rig's lighting and other electrical appliances. Again, not all rigs have an electrician.

7

Figure 13. *A mud logger monitors well conditions from his portable laboratory on the well site.*

As for personnel who work for the service and supply companies, their jobs vary according to the nature of the service they provide to the operator and contractor. For example, mud companies—companies that supply the components for the drilling fluid—almost always have a **mud engineer** (a "mud man") on duty at the rig. His duties are to formulate the mud to the specifications of the operator, see that it keeps its needed properties by running regular tests on it, and report on its properties to the company man. Also, the operating company sometimes employs a **mud logger** to stay on the well site at all times and constantly monitor what is happening downhole as the well is drilled (fig. 13).

At various stages of drilling, **casing crews** are called in to run special pipe called *casing* into the hole (fig. 14). Casing lines the hole to keep it from caving in and seals off troublesome formations that could cause

problems. Since this casing has to be cemented in place, a **cementing service company** is usually called in to perform this task. Many other services are required as the well is drilled.

Figure 14. *A casing and cementing crew, working on the rig floor, lubricates the threads on a length of casing that is about to be lowered into the hole.*

8

In summary, the people who work on today's rigs are responsible for the safe and efficient operation of millions of dollars' worth of complex equipment. The work is not always easy—the weather can be terrible, the location is often remote, the hours are long, and an element of danger exists. Nevertheless, for those who can qualify, drilling is an exciting, dynamic, and rewarding field.

WORK SHIFTS

Unlike the company man and toolpusher on land rigs, the driller, derrickman, and rotary helpers do not work around the clock. Instead, they sometimes work in 8-hour shifts called *tours* (pronounced "towers"). Since drilling goes on 24 hours a day, seven days a week, there are often four crews; however, considerable variation exists. In cases where four crews are employed, the daylight tour may work from 7 A.M. to 3 P.M., the evening tour from 3 P.M. to 11 P.M., and the morning or graveyard tour from 11 P.M. to 7 A.M. These hours can vary, depending on the individual contractor's policy, which is set according to the manpower he has available. For example, if four crews are available, three of them may work five or six days straight and get two days off. The fourth crew, often called the *relief crew,* then fills in as needed for the crews taking their days off. This too may vary. For example, in some areas (especially offshore) two crews work 12-hour tours. Also, they usually work seven days, have seven days off, work seven more days, and so on, alternating with another crew.

TRAINING

The people who work on drilling rigs come from a variety of backgrounds. The toolpusher usually attains his supervisory position only after years of experience working on rigs. Most likely he started out as a rotary helper (called a roughneck in the old days), advanced to derrickman, then to driller, and, after working as a driller for a number of years, was promoted to toolpusher when the contractor had an opening.

The new hand on a rig starts out as a rotary helper, whose first job is mainly to watch and listen. If he is lucky (and today most are), he works under a driller who patiently explains and works with him until he is able to competently perform the tasks on hand. The more experienced rotary helpers also help the newcomer learn how to safely do his job. In addition, many contractors have established training programs for new rig hands, in which practical work experience on the rig is combined with classroom work.

The International Association of Drilling Contractors (IADC), headquartered in Houston, Texas, distributes a number of training aids to assist not only the new person, but experienced workers as well. Also, several universities have services that provide petroleum training for floormen. Among these are the University of Southwestern Louisiana in Lafayette, Louisiana State University in Baton Rouge, Nicholls State at Thibodaux, Louisiana, the University of Oklahoma in Norman, the University of Houston—Victoria, Texas A&M University in College Station, and Petroleum Extension Service (PETEX) of The University of Texas at Austin. In Canada the Petroleum Industry Training Service and the Canadian Association of Oilwell Drilling Contractors (CAODC) conduct several courses designed for rig crew training (fig. 15).

Figure 15. Training plays a big part in the life of a crew member. Here the crew receives specialized training in well control techniques.

THE DRILL SITE

The drill site—the actual location on which the well is to be drilled—is selected by the operating company. The company's decision on exactly where to drill is based on several factors. The most important factor is geological; that is, the company must believe strongly that hydrocarbons (oil and/or gas) exist in the subsurface under the spot where the well will be started. This belief is based on

geological study. To determine whether or not hydrocarbons may exist is usually the job of the company's staff of geologists; or sometimes, the company will hire the services provided by a company of geologists whose specialty is finding likely sources of petroleum.

In addition to the geological factors, legal and economic factors must be considered. For

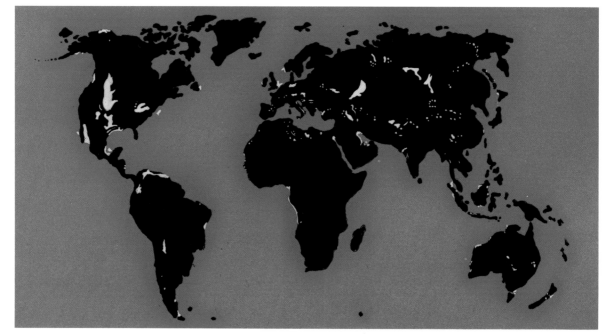

Figure 16. Oil and gas have been found in various parts of the world (indicated by white areas). Some major discoveries barely show because full-scale drilling in the area has not yet begun.

example, the company must obtain the right to drill for and produce oil and gas on the land. Also, the company must have money to not only purchase or lease the right to drill and produce on the site, but to pay for the costs of drilling. All of these factors, plus many others, have to be considered in selecting a drilling site.

Oil and gas wells are being drilled in almost every country in the world, on land, in marshes, and offshore (fig. 16). These wells are generally classified as two kinds: exploration wells and development wells.

An *exploration well*, or "wildcat," is one that is drilled primarily for the purpose of determining that oil or gas actually exists in a subsurface rock formation. Before a well is drilled, it is at best an educated guess that petroleum exists in a formation. It is only after a well is drilled into the formation that the presence of oil or gas can definitely be confirmed or denied to be present.

A *development well* is a well that is drilled after an exploration well has confirmed the presence of petroleum in the formation. Usually, it takes several development wells to efficiently produce hydrocarbons from a formation.

Major oil and gas reserves have been discovered on nearly every continent, including the continental shelves (that area of a continent relatively near the shore but covered by ocean or sea); in gulfs, bays, and marshlands; and in deserts, frozen wastelands, and tropics (fig. 17). Each discovery brings new drilling problems that have to be solved.

Figure 17. Drilling takes place in very remote areas from the tropics (as shown here) to the Arctic.

For example in the jungle and in the Arctic, drilling rigs and supplies must be brought in where no roads or highways exist. In offshore

Figure 18. Drilling on the North Slope of Alaska presents cold-weather difficulties.

areas, the threat of hurricanes and major storms is ever present. On the North Slope of Alaska, bitter cold and blizzards must be overcome (fig. 18). Sometimes rigs are located in environmentally sensitive areas (fig. 19).

As new regions change from exploration to exploitation, the emphasis changes from drilling to production. When development wells are drilled, full advantage is taken of the information gained when the exploration, or wildcat, wells were drilled. As a result, the costs of drilling are reduced as problems are solved.

Figure 19. This rig is located on a dairy farm in South Texas.

FINDING OIL AND GAS

Hydrocarbons—crude oil and natural gas—are found in certain layers of rock that are usually buried deep beneath the surface of the earth. In order for a rock layer to qualify as a good source of hydrocarbons, it must meet several criteria.

Characteristics of Reservoir Rocks

For one thing, good reservoir rocks (a reservoir is a formation that contains hydrocarbons) have *porosity*. Porosity is a measure of the openings in a rock, openings in which petroleum can exist. Even though a reservoir rock looks solid to the naked eye, a microscopic examination reveals the existence of tiny openings in the rock (fig. 20). These openings are called pores. Thus a rock with pores is said to be porous and is said to have porosity.

Another characteristic of a reservoir rock is that is must be *permeable*. That is, the pores of the rock must be connected together so that hydrocarbons can move from one pore to

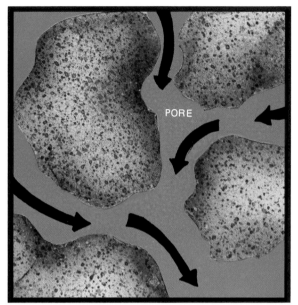

Figure 21. If pores of the rock are connected so that oil and gas move within the rock, the rock is permeable. Arrows indicate flow from pore to pore.

another (fig. 21). For unless hydrocarbons can move and flow from pore to pore, the hydrocarbons remain locked in place and cannot flow into a well.

In addition to porosity and permeability, reservoir rocks must also exist in a very special way. To understand how, it is necessary to cross the time barrier and take an imaginary trip back into the very ancient past.

Imagine standing on the shore of an ancient sea, millions of years ago. A small distance from the shore, perhaps a dinosaur crashes through a jungle of leafy tree ferns, while in the air, flying reptiles dive and soar after giant dragonflies. In contrast to the hustle and bustle on land and in the air, the surface of the sea appears very quiet. Yet, the quiet surface condition is deceptive. A look below the surface reveals that life and death occur constantly in the blue depths of the sea. Countless millions of tiny microscopic organisms eat, are eaten, and die. As they die, their small remains fall as a constant rain of organic matter that accumulates in enormous quantities on the seafloor. There, the remains are mixed in with the ooze and sand that form the ocean bottom.

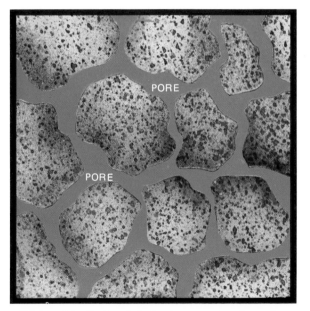

Figure 20. A simulated microscopic view of a reservoir rock reveals the openings between the individual grains that make up the rock. If oil or gas is present in the rock, it occurs in the openings called pores.

As the countless millennia march inexorably by, layer upon layer of sediments build up. Those buried the deepest undergo a transition; they are transformed into rock. Also, another transition occurs: changed by heat, by the tremendous weight and pressure of the overlying sediments, and by forces that even today are not fully understood, the organic material in the rock becomes petroleum. But the story is not over.

For, while petroleum was being formed, cataclysmic events were occurring elsewhere. Great earthquakes opened huge cracks, or faults, in the earth's crust. Layers of rock were folded upward and downward. Molten rock thrust its way upward, displacing surrounding solid beds into a variety of shapes. Vast blocks of earth were shoved upward, dropped downward, or moved laterally. Some formations were exposed to wind and water erosion and then once again buried. Gulfs and inlets were surrounded by land, and the resulting inland seas were left to evaporate in the relentless sun. Earth's very shape had been changed.

Meanwhile, the newly born hydrocarbons lay cradled in their source rocks. But as the great weight of the overlying rocks and sediments pushed downward, the petroleum was forced out of its birthplace. It began to migrate. Seeping through cracks and fissures, oozing through minute connections between the rock grains, petroleum began a journey upward. Indeed, some of it eventually reached the surface where it collected in large pools of tar, there to lie in wait for unsuspecting beasts to stumble into its sticky trap. However, some petroleum did not reach the surface. Instead, its upward migration was stopped by an impervious or impermeable layer of rock. It lay trapped far beneath the surface. It is this petroleum that today's oilmen seek.

Types of Petroleum Traps

Geologists have classified petroleum traps into two basic types: structural traps and stratigraphic traps.

Structural traps are traps that are formed because of a deformation in the rock layer that contains the hydrocarbons (fig. 22). Two

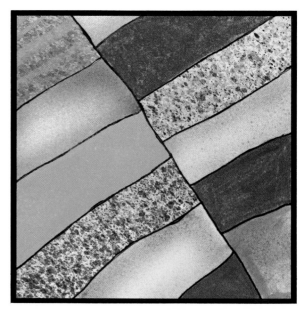

FAULT TRAP

ANTICLINAL TRAP

☐ Impermeable rock ☐ Reservoir rock

Figure 22. Two types of structural traps are the fault trap and the anticlinal trap.

common examples of structural traps are fault traps and anticlines.

A *fault trap* occurs when the formations on either side of the fault have been moved into a position that prevents further migration of petroleum. For example, an impermeable

14

formation on one side of the fault may have moved opposite the petroleum-bearing formation on the other side of the fault. Further migration of petroleum is prevented by the impermeable layer.

An *anticline* is an upward fold in the layers of rock, much like an arch in a building. Petroleum migrates into the highest part of the fold, and its escape is prevented by an overlying bed of impermeable rock.

Stratigraphic traps are traps that result when the reservoir bed is sealed by other beds or by a change in porosity or permeability within the reservoir bed itself (fig. 23). There are many different kinds of stratigraphic traps. In one type, a tilted or

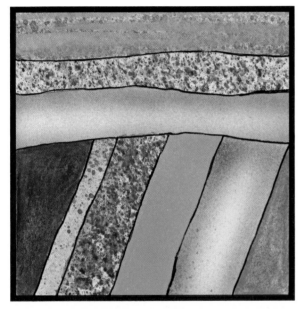

TRUNCATION

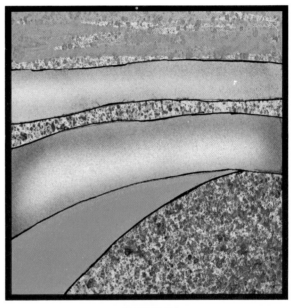

PINCH-OUT

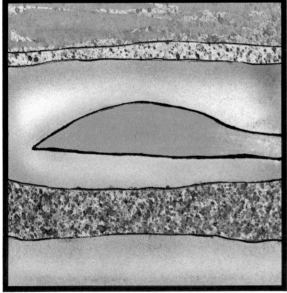

SURROUNDED

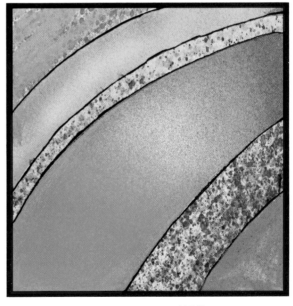

POROSITY CHANGE

 Impermeable rock

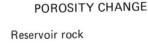 Reservoir rock

Figure 23. Stratigraphic traps

inclined layer of petroleum-bearing rock is cut off or truncated by an essentially horizontal, impermeable rock layer. Or sometimes a petroleum-bearing formation pinches out; that is, the formation is gradually cut off by an overlying layer. Another stratigraphic trap occurs when a porous and permeable reservoir bed is surrounded by impermeable rock. Still another type occurs when there is a change in porosity and permeability in the reservoir itself. The upper reaches of the reservoir may be impermeable and nonporous, while the lower part is permeable and porous and contains hydrocarbons.

Locating Petroleum Traps

Of course, it is all very well to know that various kinds of subsurface traps contain petroleum, but the real problem comes in finding such traps. In the early days of oil exploration, geologists had to content themselves with looking for features on the surface that indicated subsurface traps. Careful examination of the surface did indeed lead to many oil finds. However, most petroleum deposits lie so deeply hidden that there is no hint on the surface of that which lies below. Witness the vast stretches of flat, arid prairie in West Texas that overlie some of the most prolific oil and gas fields in the United States. Or consider that much of the world's oil and gas probably lies offshore, covered by hundreds or thousands of feet of water and by more thousands of feet of rock.

Fortunately, the science of oil exploration has advanced right along with all the other scientific discoveries of this century. In finding traps for oil, the science of concern is called *seismology.* Briefly put, seismology involves creating sound waves on the surface that penetrate downward into the rock layers (fig. 24). Each formation reflects the sound wave back to the surface, where sensitive instruments record and measure the intensity of the reflections. By carefully interpreting the measurements, exploration geologists are able to deduce the shape and extent of the subsurface formations. Since petroleum traps often have a characteristic configuration, it follows that readings taken from seismograms, or seismic sections, should reveal the possible

existence of a petroleum reservoir. However, seismology is not a perfect science. Although it is certainly better than a blind guess or a hunch, the odds are that six of every seven new, or *wildcat,* wells drilled will be dry. Only one of every eight will result in the discovery of an oil field capable of producing up to one million barrels of oil, which is a very small field. The odds against finding a large field—one capable of producing 50 million barrels of oil or more—are 3,591 to 1. In other words, only one wildcat well in 3,591 will lead to the discovery of a large field.

Figure 24. Mounted underneath this truck (above) is a vibrator, a device that creates low-frequency sound waves. A large area can be covered as the vibrator truck moves from one spot to the other. A closeup of the vibrator (below) shows it firmly pressed against the ground as the sound waves are generated. The waves travel downward and are reflected by the various formation beds through which the waves pass. Sensitive instruments called geophones pick up the sonic reflections and send them to a recorder. Interpretation of the recordings can reveal the location of favorable structures.

Figure 25. The drill site is cleared and leveled.

SELECTING THE DRILL SITE

In order for the operating company to reach the point where it can tell the drilling contractor the exact spot on which to drill, several steps have to be taken. Seismic sections are reviewed and analyzed. Lease terms and agreements are thoroughly reviewed by legal experts for clear title and right-of-way for access. Exact boundaries and locations are established or confirmed by land surveyors. The necessary funds to pay for drilling the well are confirmed, whether the funds come from corporate financing, loans, or the pocketbooks of one or more private citizens.

After evaluating the data and clearing any legal problems, the operating company selects a drilling site. On land, its hope is that the spot is at least fairly accessible, reasonably level, and situated such that minimum damage to the surrounding environment will occur as the rig is moved on location. Offshore, the hope is that the weather is reasonably good, at least most of the time, and that the water is not too terribly deep. Although many famous oil fields were discovered by guess and by gosh, the fact remains that the only sure way to prove that oil or gas lies buried under some likely spot is to probe for it with a drill. So, it is now that the operating company prepares its specifications for drilling the well and invites several contractors to bid on the job. One of them gets the contract to drill.

PREPARING THE SITE

After a contractor gets the job to drill the well on land, the location is prepared to accommodate the rig and its equipment. Offshore, buoys are set to mark the site. For an ordinary land location, the site is cleared and leveled, and access roads and a turnaround are built (fig. 25). Since the main component of drilling mud is usually water, a water well is drilled, or pumps and a waterline are installed to bring water from a suitable stream, river, or lake. Offshore, seawater is used. At a land site, an earthen pit is bulldozed out and lined with plastic (fig. 26). This reserve pit serves as a place of disposal for used or unneeded

Figure 26. Earthen reserve pits are lined with plastic sheeting to protect the soil.

17

Figure 27. Boards are laid (foreground) to permit traffic during rainy weather.

drilling mud, cuttings from the hole, and other waste that invariably accumulates around the site. The plastic lining prevents pollution of the soil. After the well is finished, the pit will be covered and leveled. Of course, if the location is in an ecologically sensitive area such as offshore or in woodlands, practically nothing will be dumped at the site. Trucks or barges are used to haul off waste material to another site where it can be safely disposed of.

Materials used to prepare the surface area and roads around the land location vary according to the terrain. Gravel or oyster shells may be used; if mud, muck, and mire are a problem, boards are laid to permit traffic during rainy weather (fig. 27). In the far north where permafrost exists, additional problems occur, and an insulating layer must be constructed so that the heat generated by the drilling equipment will not melt the permafrost, causing the rig to settle into the thawed soil. A thick layer of gravel may be laid, or polyurethane foam may be needed if gravel is scarce.

After the site is made ready, the next step is to prepare the spot where the hole will actually be drilled. Sometimes, at land locations, a rectangular pit called a *cellar* is dug (fig. 28). Later, the rig is placed directly over the cellar, which provides extra space for drilling accessories that will be installed under the rig. In the middle of the cellar, the first part of the well is actually begun but not necessarily with the big drilling rig. Instead, a

Figure 28. A cellar is dug and lined with boards.

18

Figure 29. A truck-mounted, light-duty rig drills the first part of the hole prior to moving the big rig onto the site.

Figure 30. The first part of the main hole is dug in the middle of the cellar.

relatively small, truck-mounted rig may be used to start the main hole (fig. 29). This first part of the hole is large in diameter but fairly shallow in depth (fig. 30). It is lined with large-diameter pipe called *conductor pipe*. (Sometimes, if the ground is soft, the conductor pipe is simply driven into the ground with a pile driver.)

Another hole is dug off to the side of the cellar. This hole is also lined with pipe and is called the *rathole* (fig. 31). The rathole serves as a place to temporarily store a piece of drilling equipment called the *kelly*. It is now time for the contractor to move in the drilling rig and equipment.

Figure 31. Rathole is lined with pipe (left) and conductor pipe lines the main borehole (right).

Figure 32. Rig components are moved into a remote area by helicopter.

MOVING EQUIPMENT TO THE SITE

Land Rigs

In areas that are accessible by roads and highways, trucks are used to transport the equipment to the site. Rig components are designed for portability and are easily loaded and unloaded. Granted, it can be intimidating to see (and frustrating to follow) some huge piece of equipment being moved down the highway, especially on a two-lane blacktop. (Rig trucks are built for power, not necessarily speed.) Still, in spite of its size, rig equipment is portable because it has to be. After finishing one well, the rig must be dismantled, moved, and put back together so it can begin another well, often at some point many miles away from the first.

But suppose there are no roads; suppose the drill site is in a jungle, perched precariously on the side of a mountain; or suppose the site is in the Arctic. Now what?

Since oilmen have never been known to allow something as mundane as logistics to defeat them, they have come up with a solution: helicopters and large cargo planes. A helicopter crane (fig. 32) has no difficulty in setting down components on the side of a mountain in the jungle. Likewise, a mammoth cargo plane, fitted with skis where wheels normally go, has little trouble landing equipment in the frozen wastes of the Arctic.

Offshore Rigs

Offshore, the way in which the rig is moved to the site depends on whether the well is a wildcat or a development well. If the well is a wildcat, some type of mobile offshore rig will be used. Mobile rigs include jackups, submersibles, semisubmersibles, and drill ships.

A *jackup rig* is built so that it floats when being moved to the location (fig. 33). Once the rig is at the site, its huge legs are cranked down until they contact the seafloor. Then the hull is raised or "jacked up" on the same legs to raise the hull and drilling deck above the water's surface (fig. 34). Conventional jackup rigs are limited to drilling in water depths of up to about 400 feet.

Figure 33. A jackup rig floats with its legs up as it is towed.

A *submersible rig* has hulls upon which it floats while being towed to the site. At the site, the hulls are flooded and come to rest on bottom (fig. 35). The drilling deck (sometimes called the *Texas deck*) is built on long steel columns that extend upward from the hulls. Thus, the drilling deck is well above the water's surface. Submersibles, like jackups, are limited to drilling in relatively shallow waters. Nevertheless, both provide a very stable platform from which drilling operations can be conducted.

A *semisubmersible rig* (fig. 36) is similar to a submersible in that it has two or more hulls upon which the rig floats as it is being towed to the location. (Some rigs of this type are capable of being used as either submersibles or semisubmersibles.) However, once on location the hulls are designed so that, when flooded, they do not settle to the bottom. In contrast they submerge only to a depth a little below the water's surface. In reality, a semisubmersible (or simply, "semi") floats but not on the water's surface. Where the weather can be stormy, semisubmersibles are often selected because of their excellent stability in rough, deep seas.

Figure 34. A jackup rig in position and drilling has each leg contacting the seafloor and the hull raised above the water's surface.

Figure 35. A submersible on location and drilling has its hulls resting on bottom.

Figure 36. A semisubmersible floats on the water's surface before and during towing to a drilling site (above). On location, hulls are flooded and submerged just below the water's surface as shown in this semisubmersible drilling in the North Sea (below).

Figure 37. A drill ship is usually used to drill discovery, or wildcat, wells in deep, remote offshore waters.

A *drill ship* is also a floater (fig. 37). Essentially, a drill ship is shaped just like any ocean-going ship; however, drilling equipment (and other modifications) make a drill ship distinctive. Drill ships are the most mobile of rigs and are often used to drill wildcat wells in deep, remote waters, far from land.

When a mobile rig makes a discovery (determined by drilling a well or several wells to confirm that significant quantities of oil or gas exist in a reservoir), development wells must be drilled. To drill development wells, an *offshore platform* must be built and erected at a suitable site in the waters over the reservoir (fig. 38). Platforms can be extremely large—so large that they are for the most part built on land, then floated and towed out to the site. This is the only time that a platform is mobile. For, once at the location, it is firmly pinned to the seafloor with pilings, which are driven very deep into the soil on bottom. Huge cranes, mounted on barges, are employed to hoist the drilling equipment into place on the platform. When made ready for drilling, several wells are drilled from a single platform without moving it.

Figure 38. An offshore platform is erected on the site to drill the development wells.

Inland Barge Rigs

In marshes, where "the water is too muddy to drink and too wet to plow" (that is, where the water is very shallow, and the soil is not firm enough to support a land rig), inland barge rigs are often used to drill both wildcat and development wells (fig. 39). As the name implies, the drilling equipment is installed on a flat-bottomed barge. When moved from one location to another, the barge floats and is towed by boats. When at the location, the barge is submerged to rest on bottom, and pilings are driven deep into the soil to keep it immobile.

Because rigs have to go where the oil is, and because oil is not particular as to where it resides, some method of transportation has to be devised. The method may be nothing more complex than a large flatbed truck, or it may be as intriguing as a giant helicopter with several tons of rig components dangling unceremoniously beneath it. Where there's oil or gas, there's a way to get a rig to it.

Figure 39. An inland barge rig is anchored with pilings driven into the soil.

RIGGING UP

Figure 40. *The substructure of this land rig is being raised into position.*

Figure 41. *This drilling rig has a standard derrick, which has been assembled piece by piece on the site.*

Figure 42. *A drilling rig with a mast as shown here is assembled by the manufacturer and moved to the site in sections. At first glance, the difference between a derrick and mast is not evident, but close examination reveals many differences.*

Given that somehow the contractor has managed to get his rig to the site, the next step is putting the rig together so that drilling can begin. This process is known as rigging up. For land rigs, first the *substructure*—the girderlike framework that rests on the ground right over the hole—is brought in and assembled (fig. 40). The substructure supports the mast or derrick, the miles of pipe that will be used to drill the hole, and the *drawworks*—the big machine that is used to hoist the drill pipe, or *drill string*, in and out of the hole. Sometimes, depending on the design, the engines for powering the rig machinery are also placed on the substructure.

With the substructure in place and assembled, and with the drawworks and engines in place, the next step is to get the derrick or mast up. Strictly speaking, a *derrick* is a towerlike structure that must be assembled piece by piece, requiring a rig-building crew to fasten the pieces together with bolts (fig. 41). On the other hand, a *mast* can be raised or lowered without disassembly; it is portable (fig. 42). Nevertheless, a mast is often referred to as "the derrick," even though it

really is not. Most modern rigs are equipped with masts, although a few rigs, especially offshore platforms, still have standard derricks, meaning that crews had to build them when the rigs were moved on location and will have to take them apart when the rigs are moved off location.

A mast is put into position in a special cradle on the substructure (fig. 43). At this time the mast is horizontal. Using the drawworks and wire-rope hoisting line, the mast is slowly raised to vertical (fig. 44). This can be a tricky operation. For one thing, several truck-mounted cranes are used to maneuver the mast into position in its cradle. Since a mast can be up to 200 feet long and is heavy, and its base has to be raised up on top of the substructure, the operation takes a closely coordinated effort between the crane operators and truck drivers. For another thing, anytime something as heavy as a mast is being raised, extreme care must be exercised to avoid accidents.

Meantime, other rigging-up operations continue. Steel pits, in which the drilling mud will be placed, are trucked into position and connected together. Safe stairways and walkways are installed to allow the crew access to the many components. Guardrails are erected (fig. 45). Auxiliary equipment for the supply of electricity, compressed air, and

Figure 43. The mast rests in its special cradle in the substructure before it is raised.

Figure 44. The mast is raised into vertical position.

Figure 45. Rigging-up operations include installing safe guardrails. Here, guardrails are being installed on the mud pit.

Figure 46. Huge, barge-mounted cranes are used to build an offshore platform rig.

water is put into operation. The large pumps (*mud pumps*) that will circulate the drilling fluid are put into place. Storage racks, bins, and living quarters for the toolpusher and company man are trucked in. Drill pipe, bits, mud components, wire rope, and other needed items are brought to the location. Eventually (say, from three days to a week), everything will be in readiness as rigging up is completed and drilling can begin.

Getting ready to drill offshore varies with the type of rig. For example, rigging up a platform rig is mainly a matter of assembly. Usually, a standard derrick is used, so it will

be assembled piece by piece. Living quarters, storage spaces, and other equipment are lifted by barge-mounted cranes (fig. 46).

Mobile offshore rigs require less rig-up time than platforms because most of the equipment is already in place and assembled. Floaters, such as semisubmersibles and drill ships, simply have to be anchored on location, and drilling operations can begin. Incidentally, some floaters are not anchored; instead, dynamic positioning is utilized to keep the rig on station (in a position more or less directly over the spot on the seafloor where the hole is to be drilled). Using sonar, computers, and, in some cases, artificial orbiting satellites to direct large motor-driven propellers called *thrusters*, the floater's position is maintained within a few degrees over the hole.

RIG COMPONENTS

The main function of a rotary rig is to drill a hole, or as it is known in the industry, to *make hole*. Making hole with a rotary rig requires not only qualified personnel, but a lot of equipment as well. In order to learn about the components that it takes to make hole, it is convenient to divide them into four main systems: power, hoisting, rotating, and circulating. Various components comprise the last three systems, but all require power to make them work.

Figure 47. Three gas engines are located in the substructure of this rig.

POWER SYSTEM

Practically every rig uses internal-combustion engines as its prime power source, or its *prime mover*. A rig's engines are similar to the one in a car except that rig engines are bigger, more powerful, and do not use gasoline as a fuel. Also, rigs require more than one engine to furnish the needed power. Most rig engines today are diesels, although some are still around that burn natural or liquefied gas as a fuel. Gas engines use spark plugs to ignite the fuel-air mixture in their combustion chambers to produce power (fig. 47).

Figure 48. The diesel engines in use on this rig are located on the ground, some distance away from the rig.

On the other hand, diesel engines do not have spark plugs (fig. 48). Instead, the fuel-air mixture is ignited by the heat that is generated by compression inside the engine. Anytime a gas is compressed, its temperature rises; compress it enough (as in a diesel engine), and, if it's flammable, it gets hot enough to ignite. Therefore, diesel engines are sometimes called compression-ignition engines, while gas and gasoline engines are called spark-ignition engines. Diesel engines have almost totally supplanted gas engines because diesel engines are generally less expensive to operate.

A rig, depending on its size and how deep a hole it must drill, may have from two to four engines. Naturally, the bigger the rig, the deeper it can drill and the more power it will need. Thus, the big rigs have three or four engines, all of them together developing up to 3,000 or more horsepower (fig. 49). Of course, once all this horsepower is developed, it must be sent, or transmitted, to the components to be put to work. Two common methods are used to transfer the power: electrical and mechanical.

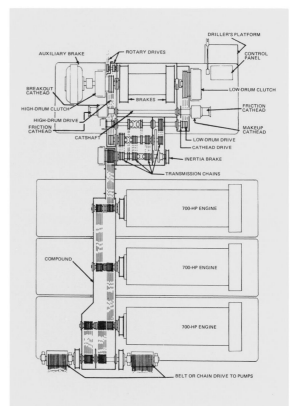

Figure 49. Multiengine and chain-drive transmission arrangement powers a mechanical drilling rig.

Mechanical Power Transmission

Up until a few years ago, most rigs were mechanical; that is, engine power was sent to the various parts of the rig by machinery such as belts and pulleys. Nowadays, diesel-electric rigs dominate the scene, but a lot of mechanical rigs are still around. On a mechanical rig, the power put out by the engines is *compounded*. In other words the two to four engines on the rig have to be joined or compounded together so all of them act as one. To do this, usually hydraulic couplings or torque converters are mounted to each engine (fig. 50). A hydraulic coupling "smooths" out the power developed by each engine, similar to the way an automatic transmission in a car smoothly transmits the power from the car's engine to its wheels.

A shaft (called an output shaft) comes out of each of the hydraulic couplings. This shaft turns because the engine is powering it. The output shafts are then mechanically linked together with pulleys and chains. The chains serve the same function as a rubber belt between two pulleys and look similar to a bicycle chain (but a lot bigger). This chain-and-pulley arrangement is known as the *compound* because it compounds or connects the power of the several engines together so that all the engine power can be used at once (fig. 51). The compound, in turn, delivers engine power through additional chain drives to the drawworks and rotary. Large belts are usually used to drive the mud pumps (fig. 52).

Figure 50. *The large radiator serves to cool the fluid in a hydraulic coupling used for transmitting power from the engines to the compound. The compound, in turn, transmits power to the drawworks, rotary, and mud pumps.*

Figure 51. *Three engines are coupled together with this chain-drive compound. The chains and pulleys comprising the compound are totally enclosed within metal guards. The back of the drawworks can be seen at extreme left.*

Figure 52. *Several belts, guarded by large steel screens, come off the compound at upper left to drive the mud pump in the foreground.*

31

Figure 53. A diesel engine (right) drives an electric generator (mounted directly to the engine) to produce electric power for the rig.

Electrical Power Transmission

Diesel-electric power is the dominant method used to drive most of today's rigs. Diesel engines, which on land rigs are usually located at ground level some distance away from the rig floor, drive large electric generators (fig. 53). The generators, in turn, produce electricity that is sent through cables to electric switch and control gear (fig. 54). From here, electricity goes through additional cables laid to electric motors that are attached directly to the equipment involved—drawworks, mud pumps, and the rotary (fig. 55).

The diesel-electric system has a number of advantages over the mechanical system. The diesel-electric system eliminates all that heavy, fairly complicated compound and chain drive. Since a compound is not needed, alignment problems are done away with; that is, the rig-up crew does not have to worry with getting a compound lined up with the engines and drawworks. Also, the engines can be placed well away from the rig floor so that engine noise for the crew is reduced.

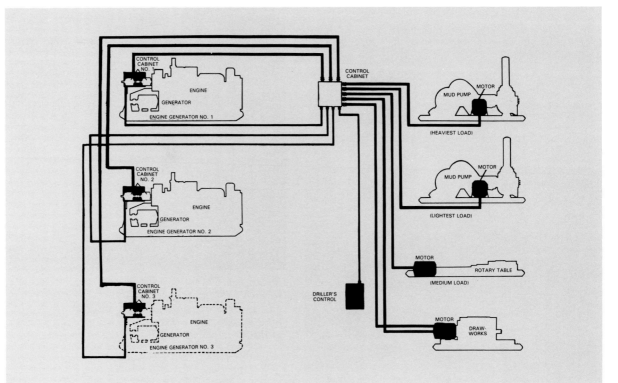

Figure 54. A diesel-electric system is one method of power transmission.

Figure 55. Electric motors drive the mud pumps. Note the blowers on top of the motors that are used for cooling the motors.

HOISTING SYSTEM

Regardless of whether the rig is mechanical or diesel-electric, its job is to make hole, and to do its job it must have a hoisting system (fig. 56). Basically, the hoisting system is made up of the drawworks (sometimes called the hoist), a mast or derrick, the crown block, the traveling block, and wire rope.

The Drawworks

The drawworks is a big, heavy piece of machinery (fig. 57). It consists of a revolving drum around which the wire rope called the *drilling line* is spooled or wrapped. It also has a catshaft on which the catheads are mounted. Further, it has several other shafts, clutches, and chain-and-gear drives for speed and direction changes. It also contains a main brake, which has the ability to stop and prevent the drum from turning. When heavy loads are being raised or lowered, the main brake is assisted by an auxiliary hydraulic or electric brake to help absorb the momentum created by a heavy load.

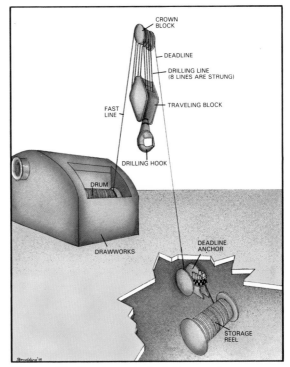

Figure 56. A rotary rig hoisting system

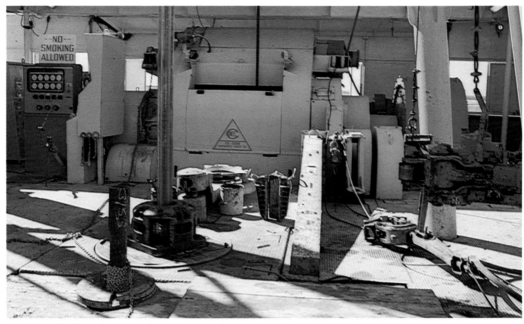

Figure 57. The drawworks consists of a revolving drum around which the wire rope is spooled.

Figure 58. The friction cathead and catline are used to hoist equipment. Note the mechanical cathead (with the chain coming out of it) mounted inboard of the friction cathead. This is the makeup cathead and is on the driller's side of the drawworks. Another mechanical cathead, the breakout cathead, is on the opposite side of the drawworks.

The Catheads

Typically, four catheads are mounted on the catshaft of the drawworks. (A cathead looks somewhat like a cat's head when viewed end on.) Two different types of catheads are mounted on each end of the catshaft, which extends out from both sides of the drawworks. On the very ends of the catshaft are the friction catheads and right next to them are the automatic or mechanical catheads.

A *friction cathead* is a relatively small, spool-shaped device. Using a large fiber rope (called the *catline*) wrapped several times around the spool of the friction cathead, a crew member can use the cathead to hoist pieces of equipment that have to be moved around on the rig floor (fig. 58). Many rigs are now equipped with small air-powered hoists that are used in place of the friction catheads. Air hoists are items of equipment separate from the drawworks and are much safer and easier to use than a cathead.

A *mechanical cathead* can be employed to make up or break out the drill string when it is being taken out or being put into the hole, or when a length of drill pipe is added as the hole deepens. The mechanical cathead located on the side of the drawworks near the driller's

Figure 59. Drilling line is kept spooled on the supply reel at the rig.

Figure 60. The large, multiple pulley is the crown block.

position is the *makeup cathead* because it plays a part when drill pipe is made up. The mechanical cathead on the other side of the drawworks is the *breakout cathead* because it plays a part when drill pipe is broken out.

The Blocks and Drilling Line

A drilling line is made of wire rope that generally ranges from 1⅛ to 1½ inches in diameter. Wire rope is similar to common fiber rope, but wire rope, as the name implies, is made out of steel wires and is a fairly complex device (fig. 59). It looks very much like what ordinary folks call "cable" but is designed especially for the heavy loads encountered on the rig.

For wire rope to be useful as the drilling line, it has to be strung up, for it arrives at the rig wrapped on a large supply reel. So, the first step in stringing up the drilling line is to take the end of the line off the supply reel and raise the end up to the very top of the mast or derrick where a large, multiple pulley is installed. This large set of pulleys is called the *crown block* (fig. 60). The pulleys are called *sheaves* (pronounced "shivs"). The drilling line is reeved (threaded) over a crown block sheave and lowered down to the rig floor. On the rig floor rests (only temporarily for now) another very large set of pulleys or sheaves called the *traveling block* (fig. 61). The end of the line is reeved through one of the traveling block sheaves, and is raised again up to the crown block. There the line is reeved over a

Figure 61. The traveling block has lines coming out of its sheaves.

Figure 62. *The deadline is wrapped around this deadline anchor and then firmly clamped. The line running out of the picture to the right goes to the supply reel.*

sheave in the crown block, lowered back down, reeved again through the traveling block, taken back up to the crown block, brought back down to the traveling block, and so on, until the correct number of lines has been strung up.

The number of lines, of course, is only one; but, since the drilling line is reeved several times over the crown block sheaves and through the traveling block sheaves, the effect is that of several lines. The number of lines strung depends on how much weight needs to be supported. For example, if a deep hole is going to be drilled, more lines are strung (say, eight or ten) than would be strung for a shallower well. (It takes more pipe and thus a heavier load to drill a deep hole than it does to drill a shallow hole.)

Once the last line has been strung over the crown block sheaves, the end of the line is lowered down to the rig floor and attached to the drum in the drawworks. Several wraps of line are then taken around the drawworks drum. The part of the drilling line running out of the drawworks up to the crown block is called the **fastline**—fast because it moves as the traveling block is raised or lowered in the derrick. The end of the line that runs from the crown block down to the wire-rope supply

reel is then secured. This part of the line is called the **deadline**—dead because, once it is secured, it does not move. Mounted in the rig substructure is a device called a *deadline anchor* (fig. 62). The deadline is firmly clamped to the anchor. Now the traveling block can be raised up off the rig floor and into the derrick by taking in line with the drawworks. To lower the traveling block, line is let out of the drawworks drum.

Crown blocks and traveling blocks usually look smaller than they actually are because of the distance from which they are seen. The sheaves around which the drilling line passes are often 5 feet or more in diameter, and the pins on which the sheaves rotate may be 1 foot or so in diameter. The number of sheaves needed on the crown block is always one more than the number needed in the traveling block. For example, a ten-line string requires six sheaves in the crown block and five in the traveling block. The extra sheave in the crown block is needed for reeving the deadline. Attachments to the traveling block include a large hook to which the equipment for suspending the drill string is attached (fig. 63).

Figure 63. *From top to bottom is the traveling block, hook, and swivel. The large hose entering the picture at right is the rotary or kelly hose.*

Masts and Derricks

Masts and derricks have to be as strong as possible yet remain portable. Consider that on a deep well the drill string may weigh as much as half a million pounds (that's 250 tons!). Yet, after finishing up one hole, the rig may be moved several miles to begin another. Manufacturers of derricks and masts usually rate their products in terms of the vertical load they can carry and the wind load they can withstand from the side. Derrick or mast capacities for vertical loads vary from 0.25 million on up to 1.5 million pounds. Most derricks and masts can withstand a wind load of 100 to 130 miles per hour. Not bad for a spindly looking collection of steel girders that has to be moved around.

ROTATING EQUIPMENT

Rotating equipment from top to bottom consists of a wondrous device known as the swivel, a short piece of pipe called the kelly, the rotary table, the drill string, and the bit. Officially, the assembly of members between the swivel and the bit, including the kelly, drill pipe, and drill collars, is termed the *drill stem*. The term *drill string* refers simply to the drill pipe; however, in the oil patch *drill string* is consistently used to mean the whole works.

The Swivel

The swivel is truly remarkable because it—
(1) sustains the weight of the drill string;
(2) permits the string to rotate; and
(3) affords a rotating, pressure-tight seal and passageway for drilling mud to be pumped down the inside of the drill string.

The swivel also has a large bail, similar to the bail or handle on a bucket but much, much larger, which fits inside the hook at the bottom of the traveling block (fig. 64). The *rotary hose* (also called the *kelly hose*) is attached to the side of the swivel. It is through this hose that drilling mud enters the swivel.

The Kelly and Rotary Table

Immediately below the swivel is attached a square (four-sided) or hexagonal (six-sided) piece of pipe called the kelly (fig. 65). Precisely why it is called the kelly is lost in the mists of

Figure 65. This four-sided kelly passes through the kelly bushing that sits inside the master bushing of the rotary.

Figure 64. Here, the hook on the bottom of the traveling block is about to be latched into the bail of the swivel.

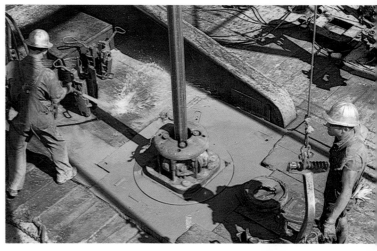

37

oil field antiquity; however, at least two persons named Kelly received patents on four-sided (square) kellys around 1920. At any rate, the kelly, like the swivel, is also a unit through which drilling mud is pumped on its way to the bottom. The reason the kelly is four- or six-sided is because it serves as a way of transferring the rotating motion of the rotary table to the drill string. The kelly fits inside a corresponding square or hexagonal opening in a device called a **kelly bushing,** which is a part of the rotary table. The kelly bushing, in turn, fits into another part of the rotary table called the **master bushing.** Thus as the master bushing rotates, the kelly bushing also rotates (fig. 66). Since the kelly mates with the kelly bushing, the kelly rotates. And finally, since

Figure 66. Turning, and thus blurred in the photo, are the master bushing and kelly bushing.

the drill pipe is connected to the bottom of the kelly, the pipe rotates when the kelly rotates. The bit also rotates because it is attached to the drill string. Incidentally, kellys are available in lengths of 40, 46, or 54 feet.

The **rotary table,** of course, is what gives rotary drilling its name. Powered either off the compound or by its own electric motor, the rotary is comprised of several parts. The master bushing drives the kelly bushing and accommodates a device called the **slips.**

Figure 67. Rotary helpers set the slips around a joint of pipe. The slips hold the pipe suspended in the master bushing of the rotary table.

A set of slips is a tapered device lined with strong, teethlike gripping elements that, when placed around drill pipe, keep pipe suspended in the hole when the kelly is disconnected (fig. 67). The kelly must be broken out, or disconnected, when a length of drill pipe is added to the drill string as the hole is drilled deeper. The kelly must also be broken out when the pipe is "tripped" (taken) in or out of the hole.

The Drill String

The drill string consists of the **drill pipe** and special, heavy-walled pipe called **drill collars** (fig. 68). Drill collars, like drill pipe, are steel

Figure 68. Dominating the picture are several drill collars; behind the collars rest many lengths of drill pipe.

Figure 69. Starting at lower right and running upward are four lengths of drill collars. Drill collars are very heavy and are used to put weight on the bit.

tubes through which mud can be pumped. Drill collars are heavier than drill pipe and are used on the bottom part of the string to put weight on the bit (fig. 69). This weight presses down on the bit to get it to drill. The operating company specifies the size and strength of the drill pipe to be used, but the type of formation being drilled and other factors determine the size and number of drill collars to be used.

A length of drill pipe is about 30 feet long, and each length is called a joint of pipe (fig. 70). Each end of each joint is threaded. One end has threads cut inside, and the other end is threaded on the outside. The inside-threaded end (the female end) is called the *box*, and the outside-threaded end (the male end) is called the *pin*. When pipe is made up

Figure 70. Joints of drill pipe are laid out on a rack at the rig prior to being run into the hole.

Figure 71. Rotary helpers stab a joint of drill pipe.

(joined together), the pin is stabbed into the box and the connection tightened (fig. 71). These threaded ends are called *tool joints*. Tool joints on drill pipe are usually welded onto the outside of the drill pipe body by a manufacturer who then cuts the threads to industry specifications.

Tool joints are not added to drill collars. The walls of drill collars are so thick that it is not necessary. Instead, the threads are cut directly onto and in drill collars. Like drill pipe, drill collars also have a box and pin (fig. 72). Thus, drill pipe can be distinguished from drill collars because drill collars do not have the bulge at either end that is characteristic of the tool joints on drill pipe (fig. 73).

Figure 73. Drill pipe is positioned on the drill floor. Note the characteristic bulge of the tool joints at the bottom of the pipe.

Figure 72. Threaded ends of drill collars are called the box (left) and pin.

40

Figure 74. Most bits have three cones; note the numerous teeth on each cone.

Figure 75. A new bit is about to be lowered into the hole.

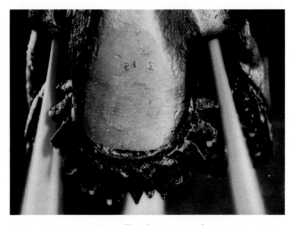

Figure 76. Drilling fluid exits with great velocity through the nozzles of the jet bit.

Figure 77. Several small industrial diamonds are set into the bottom and sides of a diamond bit.

Bits

Bits for rotary drilling are very intriguing. Years of research have gone into bit design. *Roller cone*, or *rock, bits* have cone-shaped, steel devices called cones that are free to turn as the bit rotates. Most roller cone bits have three cones (fig. 74) although some have two and some have four. Bit manufacturers either cut teeth out of the cones or insert very hard tungsten carbide buttons into the cones. The teeth are responsible for actually cutting or gouging out the formation as the bit is rotated (fig. 75). All bits have passages drilled through them to permit drilling fluid to exit.

Jet bits have nozzles that direct a high-velocity stream or jet of drilling fluid to the sides and bottom of each cone, so that rock cuttings are swept out of the way as the bit drills (fig. 76).

Diamond bits do not have cones; nor do they have teeth. Instead, several diamonds are embedded into the bottom and sides of the bit (fig. 77). Since diamonds are so hard, diamond bits are sometimes used to efficiently drill rock formations that are quite hard. They are also used to drill soft formations effectively.

41

CIRCULATION SYSTEM

Drilling Fluid

Drilling fluid—*mud*—is usually a mixture of water, clay, weighting material, and a few chemicals (fig. 78). Sometimes oil may be used instead of water, or a little oil is added to the water to give the mud certain desirable properties. Drilling mud serves several very important functions. Mud is used to raise the cuttings made by the bit and lift them to the surface for disposal. But equally important, mud also provides a means for keeping underground pressures in check. Since a hole full of drilling mud exerts pressure (just as a swimming pool full of water exerts pressure, which is why a person's ears sometimes hurt when he dives to the bottom in the deep end of a pool), the mud pressure can be used to contain pressure in a formation. The heavier or denser the mud is, the more pressure it exerts. So weighting materials—barite is the most popular—are added to the mud to make it exert as much pressure as needed to contain formation pressures (fig. 79). Clay is added to the mud so that it can keep the bit cuttings in suspension as they move up the hole. The clay also sheaths the wall of the

Figure 78. Drilling mud is kept ready for circulating in one of the pits at the rig.

hole. This thin veneer of clay called *wall cake* makes the hole stable so it will not cave in or slough (pronounced "sluff"). Numerous chemicals are available to give the mud the exact properties it needs to make it as easy as possible to drill the hole.

Figure 79. Here, the derrickman adds clay and weighting material to water to make drilling mud.

Circulating Equipment

The equipment in the circulating system consists of a large number of items (fig. 80). The **mud pump** takes in mud from the mud pits and sends it out a discharge line to a

42

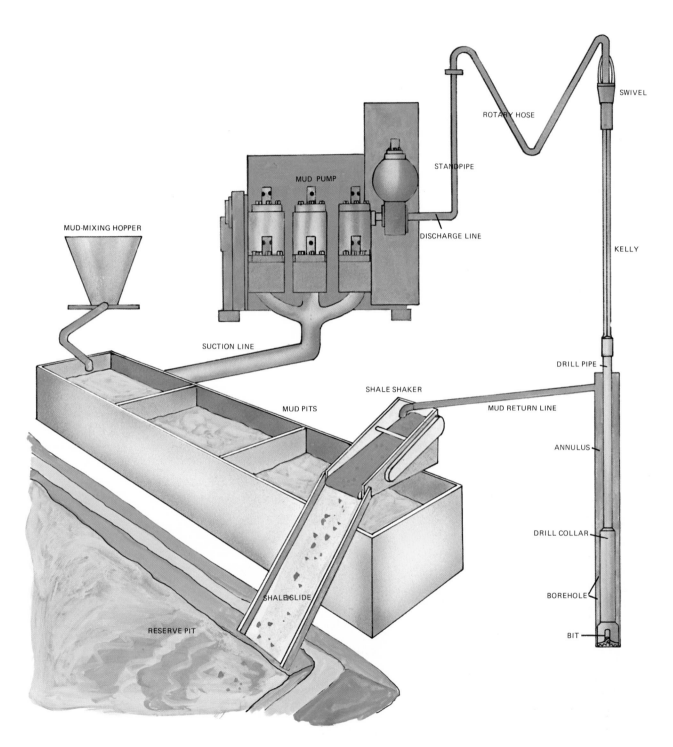

SWIVEL

ROTARY HOSE

STANDPIPE

MUD PUMP

KELLY

DISCHARGE LINE

MUD-MIXING HOPPER

SUCTION LINE

DRILL PIPE

SHALE SHAKER

MUD PITS

MUD RETURN LINE

ANNULUS

DRILL COLLAR

SHALE SLIDE

BOREHOLE

RESERVE PIT

BIT

Figure 80. Drawn without the derrick, this diagram shows the relationship of the many components of the circulating system.

Figure 81. The mud pump takes mud from the pits and discharges it into the drill string. Most rigs have two pumps but usually only one at a time is used.

standpipe (fig. 81). The standpipe is a steel pipe mounted vertically on one leg of the mast or derrick. The mud is pumped up the standpipe and into a flexible, very strong, reinforced rubber hose called the ***rotary hose***, or ***kelly hose*** (fig. 82). The rotary hose is connected to the swivel. The mud enters the swivel; goes down the kelly, drill pipe, and drill collars; and exits at the bit. It then does a sharp U-turn and heads back up the hole in the ***annulus***. The annulus is the space between the outside of the drill string and wall of the hole.

Figure 82. A rotary hose loops between the swivel at the left and the standpipe at right.

Finally, the mud leaves the hole through a steel pipe called the **mud return line** and falls over a vibrating, screenlike device called the **shale shaker** (fig. 83). On a land rig, the shaker screens out the cuttings and dumps them into one of the reserve pits (the earthen pits excavated when the site was being prepared). On offshore rigs, the shaker also screens out the cuttings, but the cuttings are dumped into a barge to be transported to a land site for disposal. In either case, the mud drains back into the mud pits and is recycled back down the hole by the mud pump. The circulating system is essentially a closed system. The mud is circulated over and over again throughout the drilling of the well. Of course, from time to time a few additions of water, clay, or other chemicals may have to be added to make up for losses or to adjust the mud's properties as new and different formations are encountered.

Auxiliary Equipment

Exacting requirements related to the maintenance of drilling mud make for some important auxiliary equipment in the system (fig. 84). For example, **agitators** installed on

Figure 83. Crew members check out the shale shaker. The large pipe running from the bottom of the picture to the shaker is the mud return line.

Figure 84. A degasser, desilter, and desander (left to right) are usually located so that the mud passes through them after it falls through the shale shaker (far right).

Figure 85. An agitator (the shaft submerged in the drilling mud) has paddles that turn to keep the mixture uniform.

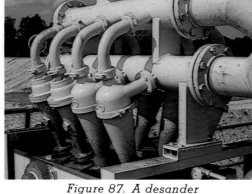

Figure 86. A desilter

Figure 87. A desander

Figure 88. A degasser

the mud pits help maintain a uniform mixture of liquids and solids in the mud (fig. 85). If any fine silt or sand is being drilled, then devices called *desilters* (fig. 86) and *desanders* (fig. 87) may be added. Since the shale shaker screen is not fine enough to remove very small particles, the mud may be sent through the desilters and desanders, which are mounted on the mud pits, so that these "fines" can be removed. Often, it is not desirable to recirculate fines back down the hole because they can erode the drill string and other components, and they can make the mud heavier than desired.

Another auxiliary in the mud system is a device called a *degasser* (fig. 88). Sometimes, small amounts of gas in a formation will enter the mud as it is being circulated downhole. It

is usually not desirable to recirculate this gas-cut mud back into the hole because the gas makes the mud lighter or less dense. If the gas is not removed by a degasser, then the mud could become so light as to allow pressure in the formation to enter the hole, and a catastrophic event known as a *blowout* could occur. (The Spindletop well was a blowout.)

Still other components in the circulating system include a *hopper*, which is a big, funnel-shaped piece of equipment used when adding solid materials like clay and barite to the mud in the pits. A *mud house* is used for storage of sacks of mud materials to keep them dry and out of the way until needed. Large, bulk-storage bins also hold great quantities of material needed in making up the mud.

46

Well Control

The drilling fluid that courses through the system also provides the first line of defense against blowouts. A blowout can be an impressive sight (fig. 89). Gas, oil, or salt water spews into the air with a tremendous roar. If gas is present, the whole thing will probably be on fire, and the rig will lie as a melted, twisted mass of junk. Human lives are threatened; pollution may occur; precious oil or gas are wasted; and a rig worth many thousands or even millions of dollars may be a total loss. Obviously, it is very desirable not to allow blowouts, and, in fact, not many occur. But, because a blowout is often a spectacular show and human lives are sometimes lost, a blowout often becomes a media event. Unfortunately, the impression may linger that blowouts are not the rarity they actually are. In reality, rig crews go to great lengths to see that the well they are drilling remains under control and does not get away from them.

A hole full of mud that weighs the right amount, or has the proper density, will not blow out. But sometimes the unexpected occurs. Since a rig crew is only human, an error can be made and formation fluids such as gas, salt water, oil, or all three enter the hole. Anytime formation fluid gets into the hole, the crew has a *kick* on its hands. When a kick occurs, it makes its presence known by certain things that happen in the circulating system. For example, the level of mud in the pits may rise above the normal level, or mud may flow out of the well even with the pump stopped, or shut down. An alert crew can spot these anomalies (even though the anomalies are sometimes subtle) and take action to prevent a blowout. It is here that the second line of defense against blowouts is brought into play—the **blowout preventers (BOPs)**.

Figure 89. A blowout and fire on an offshore drilling platform

On land rigs and on offshore rigs, such as platforms and jackups that are not floaters, BOPs are attached to the top of the well beneath the rig floor (fig. 90). The preventers are nothing more than large, high-pressure valves capable of being remotely controlled. When closed, they form a pressure-tight seal at the top of the well and prevent the escape of fluids. On floating offshore rigs, such as semisubmersibles and drill ships, the blowout preventers are attached to the well on the seafloor (fig. 91).

Two basic types of blowout preventers are annular and ram. The *annular preventer* is usually mounted at the very top of the stack of BOPs. It is called an annular because it seals off the annulus between the drill pipe or kelly and the side of the hole. An annular BOP can also seal off an open hole—a hole that has no pipe in it. Below the annular preventer in the BOP stack are typically mounted two, three, or even four ram-type BOPs. *Ram preventers* get their name from the fact that the devices that seal off the well are large, rubber-faced blocks of steel that are rammed together, much like a couple of fighting rams butting heads. Of the ram-type preventers there are *blind rams*, which seal off open hole, and *pipe rams*, which seal off the hole when drill pipe is in use. Usually, only the annular preventer will be closed when the well kicks, but should it fail, or should it be necessary to use special techniques, the ram-type preventers are used as a backup.

Figure 90. Blowout preventer stack on a land rig. At the top of the stack is the annular preventer. Two ram preventers are nippled up below the annular. Note the vertical pipe running from top to bottom—this is the mousehole.

Figure 92. A choke manifold, a series of special, remote-controlled valves, enables the driller to relieve downhole pressures without loss of control when the well is closed in by the blowout preventer.

Of course, closing in the well with one or more of the blowout preventers is only the first step that must be taken. In order to resume drilling, the kick must be circulated out and mud of the proper weight circulated in. Therefore, a series of valves called the **choke manifold** is installed as part of the system (fig. 92). A choke is simply a valve whose opening is capable of being restricted or pinched in. It can be fully closed or fully open, and it may be infinitely variable in size between open and closed. In order to circulate the kick out of the well and pump heavy mud in, the choke is fully opened, the mud pump started, and, as the kick starts moving up the hole, the choke opening reduced in size by an amount that holds just enough back-pressure to allow the mud and kick out but prevents further entry of formation fluid. Once the kick is out and the heavier mud in, a few checks are made, and normal drilling operations can resume.

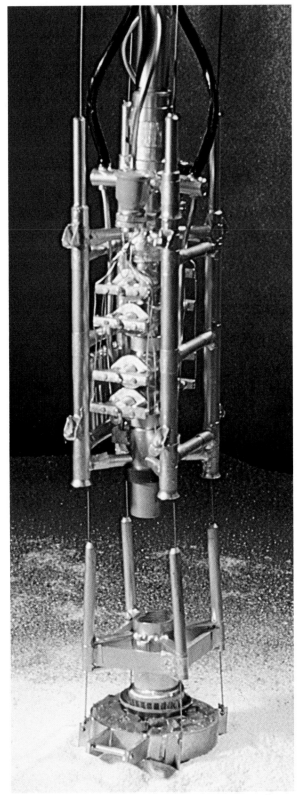

Figure 91. A subsea blowout preventer stack is lowered to the wellhead on the seafloor.

DRILLING OPERATIONS

Normal drilling operations are, after all, what the rig and crew are mainly hired to do. Simply stated, that which constitutes normal drilling operations are (1) keeping a sharp bit on bottom, drilling as efficiently as possible; (2) adding a new joint of pipe as the hole deepens; (3) tripping the drill string out of the hole to put on a new bit and running it back to bottom, or making a round trip; and (4) running and cementing casing, large-diameter steel pipe that is put into the hole at various, predetermined intervals. Often, however, special casing crews are hired to run the casing and usually a cementing company is called on to place cement around the casing to bond it in place in the hole. Still, the rig crew usually assists in casing and cementing operations.

DRILLING THE SURFACE HOLE

To get the ball rolling, assume that the crew is ready to begin drilling the first part of the hole. Remember that about 20 to 100 feet has already been started and lined with conductor pipe as described in the section on preparing the drilling site. The diameter of the conductor pipe varies, of course, but in this example assume it is 20 inches. Therefore, the first bit used will have to be smaller than 20 inches. Here, a 17½-inch bit is chosen. This fairly large bit is made up on the end of the first drill collar, and both bit and drill collar are lowered into the conductor hole. Enough collars and drill pipe are made up and lowered in until the bit is almost to bottom.

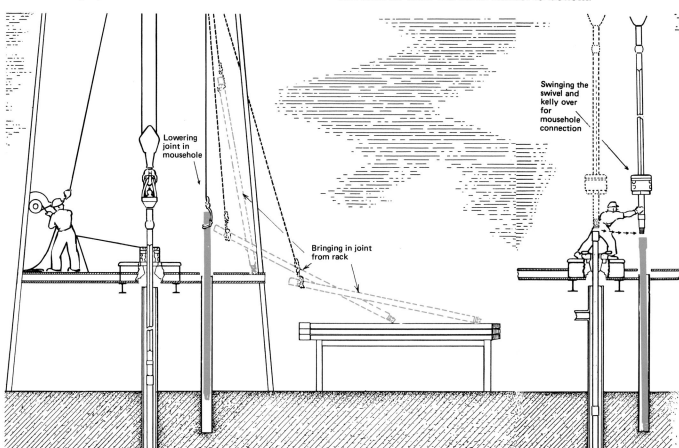

Lowering joint in mousehole

Bringing in joint from rack

Swinging the swivel and kelly over for mousehole connection

Then the kelly is picked up out of the rathole and made up on the top joint of drill pipe sticking out of the rotary table. The drill string is suspended in the rotary table by the slips. With the kelly made up, the driller starts the mud pump, or *breaks circulation,* lowers the kelly bushing to engage the master bushing in the rotary table, and begins rotating the drill stem and bit.

Next, he slowly lowers the drill stem until the bit tags (touches) bottom and begins making hole. Using a special instrument—a *weight indicator*—the driller monitors the amount of weight put on the bit by the drill collars. Weight presses the bit firmly onto the bottom so that it can penetrate the rock.

When the kelly is *drilled down*—when the hole is deepened by an amount equal to the kelly's length—the driller stops the rotary, raises the drill stem off bottom, and shuts down the mud pump. The floor crew sets the slips and swings two big wrenches called *tongs* into action. They latch the *breakout tongs* on the kelly and the *backup tongs* on the drill pipe. The driller then engages the breakout cathead, and the cathead reels in the *tong pull line,* which is attached to the end of the breakout tongs. The cathead pulls on the line with tremendous force and causes the breakout tongs to loosen, or break out, the kelly from the drill pipe. The backup tongs keep the drill pipe from turning as force is applied to the breakout tongs. With the kelly loosened, the crew removes the tongs, and the kelly is spun out (unscrewed) from the drill pipe. Often, a *kelly spinner,* an air-actuated device mounted near the top of the kelly, is used to spin out the kelly.

Meanwhile, the crew latches the backup tongs onto a joint of pipe in the *mousehole.* The mousehole is a hole in the rig floor into which a length of large-diameter pipe is placed. A joint of drill pipe is put inside the large-diameter pipe—the mousehole—prior to the joint of drill pipe being made up in the drill string (figs. 93 and 94). With the backup tongs latched, the crew swings the kelly over the pipe in the mousehole and stabs the kelly into the drill pipe. The kelly is spun up. The crew then latches the *makeup tongs* on the kelly, and the kelly is made up to final tightness. The crew unlatches the tongs, and the driller raises the kelly and drill pipe joint out of the

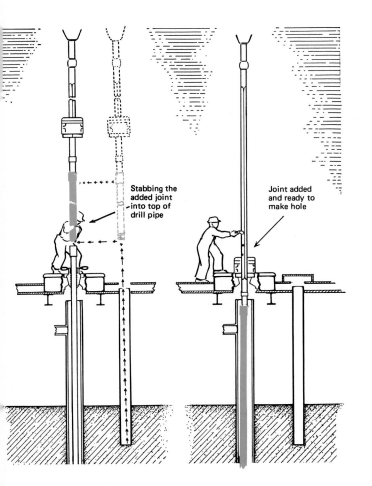

Stabbing the added joint into top of drill pipe

Joint added and ready to make hole

Figure 93. The mousehole holds a length of drill pipe while it is being connected to the kelly. The new joint of pipe is then stabbed into the top of the drill pipe in the hole.

1

2

3

Figure 94. Mousehole connection

1. When the top of the kelly nears the rotary (at right), another joint of pipe must be added to drill deeper. The new length of drill pipe has been placed in the mousehole (at left).

2. Floormen break out the kelly so that it can be swung over to the joint of pipe in the mousehole.

3. The kelly is made up on the joint of pipe in the mousehole and is bucked up with the tongs.

4. The new joint of pipe is picked up, stabbed, and spun up on the pipe hanging in the rotary, then tonged up tight before being lowered into the hole to drill another 40 or so feet.

4

mousehole. The crew swings the assembly over the drill pipe hanging in the hole and stabs the new joint into the drill stem. The two joints are then spun up, and tongs are used to buck up the new joint to final tightness. Finally, the slips are pulled, the pump started, the bit placed back on bottom, and drilling continued. What has been described is *making a connection;* it takes place each time the kelly is drilled down. When drilling near the surface, where drilling is usually easy, the crew will probably make several connections while they are on tour.

At some predetermined depth, perhaps as shallow as a few hundred feet to as deep as two or three thousand feet, drilling stops. The drilling stops because this first part of the hole—the *surface hole*—is drilled only deep enough to get past soft, sticky formations, gravel beds, freshwater-bearing formations, and so forth that lie relatively near the surface. At this point, the drill string and bit are tripped out of the hole.

TRIPPING OUT

To trip out, the slips are set, and the kelly is broken out and set back in the rathole (figs. 95 and 96). Also, the swivel is removed from the hook at the bottom of the traveling block. So, stored in the rathole are the kelly, kelly

bushing, swivel, and attached rotary hose. Still attached to the bottom of the hook, however, is a set of *elevators.* (They have been there all the time; they just aren't needed while the bit is on bottom drilling.) The elevators are a set of clamps that the floormen latch onto the drill pipe to allow the driller to raise or lower the drill string out of or into the hole when a trip is being made. The driller lowers the traveling block and elevators down to the point where the crew can latch the elevators onto the pipe. The driller raises the traveling block, thus raising the elevators and pipe, and the floormen remove the slips. Then they break out the pipe, carefully swing the lower end of it off to one side, and set it down on the rig floor.

Meanwhile, the derrickman, using a safety harness and climbing device, has climbed up into the mast or derrick to his position on the

Figure 95. The kelly is stored in the rathole while the drill pipe is tripped out of the hole.

53

onkeyboard, a small working platform
ovided for him. As the top of the pipe
aches the derrickman's position, he guides it
ck into a rack called the *fingerboard.* The
gerboard has steel protrusions, or fingers,
mewhat like the tines on a fork, into which
e pipe is set. When the floor crew sets the
ttom end of the pipe onto the rig floor, the
rrickman unlatches the elevators. The driller
en lowers the elevators down to floor level,
d the crew latches them onto the next length
pipe to be pulled out of the hole.

Not every single joint of drill pipe or drill
lar is broken out one at a time. Instead, the
ng is most often pulled three joints at a
e. So even though the pipe was put into the
e one joint at a time when drilling was
ing on, it is pulled out three joints at a time.
ree joints of pipe connected together
nstitute what is termed a *stand.* If three
nts comprise a stand, and that is the usual
se, then the stand is sometimes called a
ibble (although this term is not heard too
ch anymore). If two joints make up a stand,
e stand is a *double.* Similarly, if four joints
e in a stand, the stand is a *fourble.*

At any rate, the driller, derrickman, and
ormen all must work together as a closely
ordinated team to trip the pipe out. Since
e surface hole is relatively shallow, it is not
long before all the drill string and bit are
of the hole.

Figure 96. Tripping out

1. The kelly is usually drilled down before the drill string is pulled out of the hole. The kelly, rotary bushing, kelly cock, and swivel are stored in the rathole while the drill pipe is tripped out and back in. Also, note the kelly spinner—the disc-shaped device with two small pneumatic hoses running to it. It is used to spin the kelly up and out during connections.

2. With the kelly out of the way, the elevators that are a part of the hook-and-block assembly are latched around the pipe just below the tool joint box. The tool joint box provides a shoulder for the elevators to pull against.

3. Rotary helpers pull the slips, and a stand is pulled from the hole.

4. When the correct amount of pipe has been pulled to make a stand, the floormen drop tapered slips into the bowl of the rotary table to wedge and support the rest of the pipe in the hole.

4

7

8

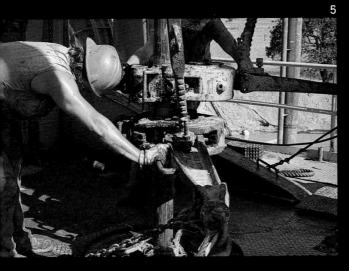

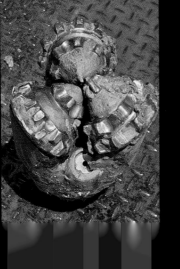

5. The connection is loosened, or broken out, with the breakout tongs (top set). Backup tongs (bottom set) keep the pipe from turning as the joint is broken out.

6. The stand is started on its swing to the place where it will be set down on the rig floor.

7. The top of the stand is pulled past the derrickman who places a rope around the pipe and racks it in the fingers to his right.

8. Floormen place stands on the rig floor in an orderly fashion.

9. The drill collars and bit are the last to come out of the hole. The bit breaker, a box specially designed for the size of bit in use, is in place in the rotary table, and the breakout tongs are used to loosen the bit. The bit is then removed from the collar by spinning it out by hand.

10. A visual inspection of the bit shows that the teeth are worn. The bit will be replaced with a new one.

Figure 97. Casing threads must be cleaned and inspected before being run into the hole.

RUNNING SURFACE CASING

Once the pipe is out, the casing crew moves in to do its work (figs. 97–102). Since this is surface hole, the first string of casing they run is called *surface casing*. Surface casing is large in diameter and, like all casing, is nothing more than steel pipe. Running casing into the hole is very similar to running drill pipe, except that the casing diameter is much larger and thus requires special elevators, tongs, and slips to fit it.

Figure 99. Casing is picked up and raised into the derrick a joint at a time.

Figure 98. Casing is placed on racks prior to being picked up.

Figure 100. Standing on a platform above the heavy-duty slips used to suspend the casing string in the hole, crew members unlatch the casing elevators. Usually, the casing crew furnishes its own casing tongs, elevators, and slips.

Figure 101. Hydraulic casing tongs are used to spin up and tighten each joint. The dial at right indicates the proper makeup.

Figure 102. The derrickman guides the casing elevators around the top of a joint of casing to lower the joint into the hole.

Also, devices called *centralizers* and *scratchers* are often installed on the outside of the casing before it is lowered into the hole (fig. 103). Centralizers·are attached to the casing and, since they have a bowed-spring arrangement, keep the casing centered in the hole after it's lowered in. Centralized casing can make for a better cement job later. Scratchers also come into play when the casing is cemented. The idea is that if the casing is moved up and down or rotated (depending on scratcher design), the scratchers will remove the wall cake formed by the drilling mud and the cement will thus be able to bond better to the hole.

Other casing accessories include a *guide shoe*, a heavy steel and concrete piece attached to the bottommost joint of casing that helps guide the casing past small ledges or debris in the hole; and a *float collar*, a device with a valve installed in the casing string two or three joints from bottom. A float collar is designed to serve as a receptacle for *cement plugs* and to keep drilling mud in the hole from entering the casing. Just as a ship floats in water, casing floats in a hole full of mud (if mud is kept out of the casing). This buoyant effect helps relieve some of the weight carried on the mast or derrick as the long string of heavy casing hangs suspended in the hole.

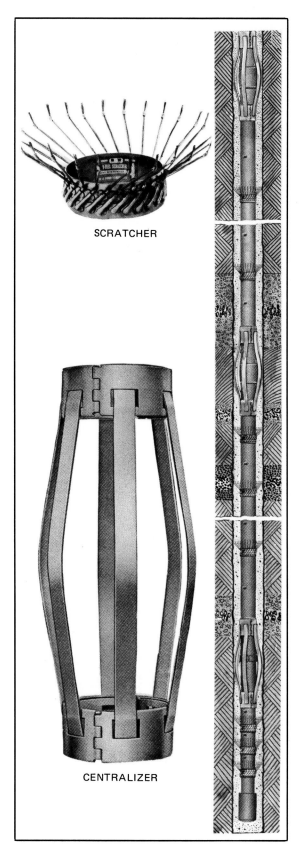

SCRATCHER

CENTRALIZER

Figure 103. An array of scratchers and centralizers is installed on the joints of casing to aid in well cementing.

Figure 104. Dry cement components are stored in large tanks ready to be transported by trucks to a land job.

Figure 105. A marine service cementing unit is used for cementing wells drilled on some platform and inland barge rigs.

CEMENTING

After the casing string is run, the next task is cementing the casing in place. An oilwell cementing service company is usually called in for this job, although, as when casing is run, the rig crew is available to lend assistance.

Cementing service companies stock various types of cement and have special transport equipment to handle this material in bulk (figs. 104 and 105). Bulk cement storage and handling equipment is moved out to the rig, making it possible to mix large quantities of cement at the site (fig. 106). The cementing crew mixes the dry cement with water, using a device called a jet-mixing hopper (figs. 107 and 108). The dry cement is gradually added to the hopper, and a jet of water thoroughly mixes with the cement to make a *slurry* (very thin, watery cement). Also used to mix water and cement is a *recirculating mixer*, frequently called an RCM. In a recirculating mixer, dry cement is wetted and mixed with water, and this wet cement is then mixed with recirculated slurry. Recirculating slurry through cement and water gives a smoother, more homogeneous slurry than a jet-mixing hopper, so RCMs are being used more and more.

Special pumps pick up the cement slurry and send it up to a valve called a *cementing head* (also called a *plug container*) mounted

Figure 106. Mobile equipment is brought in to cement a land well.

59

Figure 107. Some of the equipment used in a cement job includes the jet mixing hopper (at lower left) into which dry cement falls to be mixed with water to make slurry. The truck-mounted pump (at right) picks up the slurry and pipes it into the casing.

on the topmost joint of casing that is hanging in the mast or derrick a little above the rig floor. Just before the cement slurry arrives, a rubber plug (called the *bottom plug*) is released from the cementing head and precedes the slurry down the inside of the casing. The bottom plug stops or "seats" in the float collar, but continued pressure from the cement pumps opens a passageway through the bottom plug. Thus, the cement slurry passes through the bottom plug and continues on down the casing. The slurry then flows out through the opening in the guide shoe and starts up the annular space between the outside of the casing and the wall of the hole. Pumping continues and the cement slurry fills the annular space.

Figure 108. Cement-handling equipment, such as this skid-mounted pump, is permanently installed on many offshore rigs.

A *top plug,* which is similar to the bottom plug except that it is solid, is released as the last of the cement slurry enters the casing. The top plug follows the remaining slurry down the casing as displacement fluid (usually salt water or drilling mud) is pumped in behind the top plug. Meanwhile, most of the cement slurry flows out of the casing and into the annular space. By the time the top plug seats on or "bumps" the bottom plug in the float collar, which signals the cementing pump operator to shut down the pumps, the cement is only in the casing below the float collar and in the annular space. Most of the casing is full of displacement fluid (fig. 109).

After the cement is run, a waiting time is allotted to allow the slurry to harden. This period of time is referred to as *waiting on cement* or simply *WOC.*

After the cement hardens, tests may be run to ensure a good cement job, for cement is very important. Cement supports the casing, so the cement should completely surround the casing; this is where centralizers on the casing help. If the casing is centered in the hole, a cement sheath should completely envelop the casing. Also, cement seals off formations to prevent fluids from one formation migrating up or down the hole and polluting the fluids in another formation. For example, cement can protect a freshwater formation (that perhaps a nearby town is using as its drinking water supply) from saltwater contamination. Further, cement protects the casing from the corrosive effects that formation fluids (as salt water) may have on it. After the cement hardens and tests indicate that the job is good, the rig crew attaches or *nipples up* the blowout preventer stack to the top of the casing. The BOP stack is pressure-tested, and drilling is resumed.

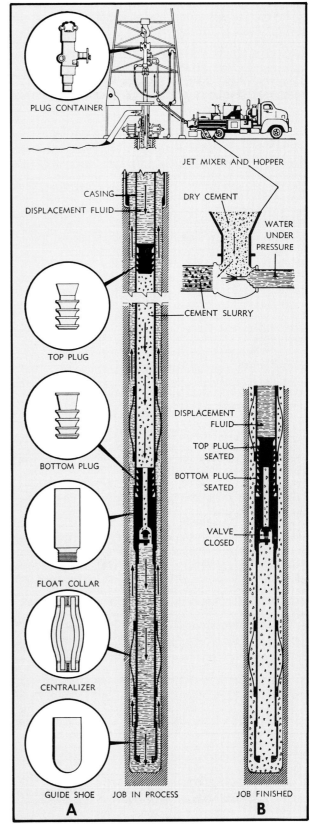

Figure 109. Cementing the casing: **A,** *the job in progress;* **B,** *the finished job.*

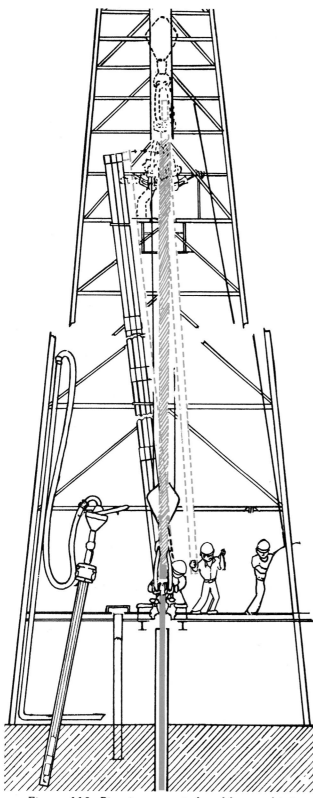

Figure 110. Pipe is connected and lowered into the hole as the kelly rests in the rathole during tripping in.

TRIPPING IN

To resume drilling, a smaller bit is selected, because it must pass down inside the surface casing. To drill the surface hole, the example rig crew used a 17½-inch bit, whereas a 12¼-inch bit will now be used. In this case, the inside diameter of the surface casing is 13⅜ inches, so in order to get adequate clearance, a 12¼-inch bit is used. As before, the bit is made up on the drill collars followed by drill pipe (fig. 110).

Tripping the pipe back into the hole calls for a special skill on the part of one of the floormen, for he will perform a job known as *throwing the spinning chain* (fig. 111). As a stand of pipe hangs suspended in the hole, one end of the spinning chain is wrapped neatly around the tool joint of the suspended pipe. The other end of the chain is connected to the makeup cathead. After the chain is wrapped, the crew stabs the next stand of pipe into the suspended stand, and the spinning chain is thrown. That is, the crewman, with a deft toss of the wrist, causes the chain to unwrap from the suspended pipe, move upward, and coil neatly around the tool joint of the stand that was just stabbed.

The driller then engages the cathead to pull on the spinning chain. As the chain is pulled off the pipe, it causes the stand of the pipe to rotate, or spin. This motion screws the spinning stand into the suspended stand. The tongs are then used to buck up the stand to final tightness. Each time a stand is made up, the spinning chain and tongs are used until all of the drill string is back in the hole.

Many contractors now use special spinning tongs to spin up pipe. Spinning tongs are air-operated and automatically spin up the stand after they are latched on the pipe, eliminating the need to throw a spinning chain.

Once the drill string is back in the hole, the cement remaining inside the lower part of the casing and the concrete inside the guide shoe is drilled out, and the next portion of the hole is continued. In this part of the hole, connections may not be as frequent because some of the formations may be hard and the drilling slower. Also, at some point the bit will get dull and have to be replaced with a new one. When this happens, it is time for a round

1

2

3

Figure 111. *Tripping in*

1. A stand of pipe is guided (stabbed) into a stand suspended in the rotary table.

2. After the stand is stabbed, one of the floormen throws the spinning chain.

3. The driller actuates the makeup cathead, which pulls the chain off the pipe causing it to spin until the tool joint shoulders meet. Continued pull on the chain, which is attached to the end of the tongs, bucks up the joint to final tightness.

4. The crew on this rig uses a set of power tongs, which both spin up and buck up the pipe and eliminate the need for a chain.

5. When the joint is tight, the slips are pulled and the pipe is lowered into the hole.

6. As the top of the stand nears the rotary, the slips are set and the elevators are unlatched and hoisted into the derrick for the next stand.

4

5

6

trip. Using the same techniques and tools described earlier, the drill string and bit are tripped out, a new bit, suitable for the type of formations being drilled, is made up, and the whole assembly of bit, drill collars, and drill pipe is tripped back in. Several round trips may be necessary before drilling is once again brought to a halt.

RUNNING AND CEMENTING INTERMEDIATE CASING

At this point, particularly in deep wells, another, smaller-in-diameter string of casing may be set and cemented in the hole. This casing string is the *intermediate string.* It runs all the way from the surface, down through the surface string, and to the bottom of the intermediate hole. Sometimes intermediate string is needed in deeper holes because almost invariably so-called troublesome formations are encountered in the hole. Troublesome formations are those that may contain formation fluids under high pressure and, if not sealed off by casing and cement, could blow out, making it difficult if not impossible to eventually produce oil or gas from the well. Or perhaps there is a *sloughing shale,* a formation composed of rock called shale that swells up when contacted by the drilling mud and falls or sloughs off into the hole. Many types of troublesome formations can be overcome while they are being drilled but are better cased off and cemented when the final portion of the hole is drilled.

DRILLING TO FINAL DEPTH

Whether intermediate casing is set or not, the final part of the hole is what the operating company hopes will be the production hole. To drill it, the crew makes up a still smaller bit, say one of 8¾ inches for this model well. This bit is tripped in, drills out the intermediate casing shoe, and heads toward what everyone hopes is pay dirt—a formation capable of producing enough oil and gas to make it economically feasible for the operating company to complete the well. Once again several bits will be dulled and several round trips will be made, but before long the *formation of interest* (the *pay zone,*

the *oil sand,* or the formation that is supposed to contain hydrocarbons) is penetrated by the hole. It is now time for a big decision. The question is, "Does this well contain enough oil or gas to make it worthwhile to run the final production string of casing and complete the well?"

EVALUATING FORMATIONS

Examining Cuttings

To help the operator make his decision, several techniques have been developed. One thing that helps indicate whether hydrocarbons have been tapped is a thorough examination of the cuttings brought up by the bit (fig. 112). The mud logger (Remember him? He's been there all along, monitoring downhole conditions in his trailer at the location.) catches cuttings at the shale shaker and by using a microscope or ultraviolet light can see whether oil is in the cuttings (fig. 113). Or he may use a gas-detection instrument.

Figure 112. Cuttings made by the bit are examined by the mud logger.

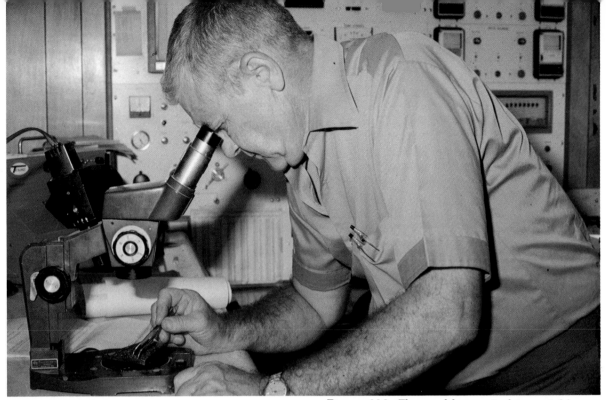

Figure 113. The mud logger in his portable laboratory at the well site views cuttings from the well under a microscope.

Figure 114. A well logging truck is brought to the well site while the drill string is tripped out.

Well Logging

Another valuable technique is well logging. A logging company is called to the well while the crew trips out all the drill string. Using a portable laboratory, truck-mounted for land rigs and permanently mounted on offshore rigs, the well loggers lower devices called logging tools into the well on wireline (fig.114). The tools are lowered all the way to bottom and then reeled slowly back up. As the tools come back up the hole, they are able to measure the properties of the formations they pass.

Electric logs measure and record natural and induced electricity in formations. Some logs ping formations with sound and measure and record sound reactions. Radioactivity logs measure and record the effects of natural and induced radiation in the formations. These are only a few of many types of logs available.

Since all the logging tools make a record, which resembles a graph or an electrocardiogram (EKG), the records, or logs (fig. 115), can be studied and interpreted by an experienced geologist or engineer to indicate not only the existence of oil or gas, but also how much may be there (fig. 116). Computers have made the interpretation of logs much easier.

INDUCTION—ELECTRICAL LOG

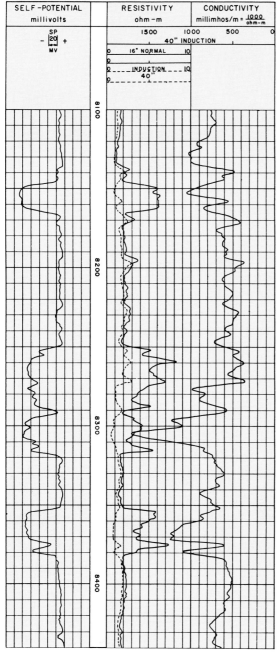

Figure 115. An electrical log

Figure 116. Examining logs at the location may show whether the well is a bonanza or a duster!

Drill Stem Testing

Still another helpful technique is the drill stem test (DST) tool (fig. 117). This tool is made up on the drill string (the drill stem) and set down on the bottom of the hole. A *packer*, an expandable hard-rubber sealing element, seals off the hole below it by expanding when weight is set down on it. A valve is opened, and any formation pressure and fluids present enter the tool. A recorder in the tool makes a graph of the formation pressure (fig. 118). Then the packer is released and the tool retrieved back to the surface. By looking at the record of the downhole pressures, a good indication of the characteristics of the reservoir can be obtained.

66

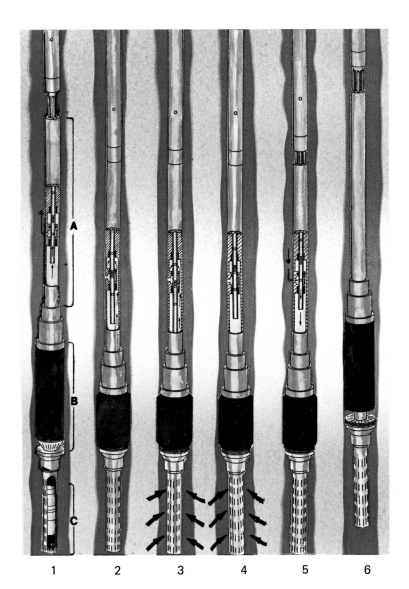

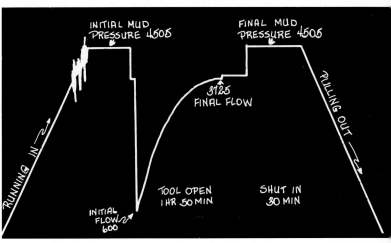

Figure 117. Principles of drill stem testing: **A,** the test valve; **B,** the packer; and **C,** pressure recorders. (1) The tool is run in on empty drill pipe, with bypass valve open; (2) packer is set on bottom to isolate the test zone; (3) production valve opens, and formation fluids enter the drill pipe; (4) test ends, and production valve closes; (5) bypass valve opens; and (6) packer recedes, and tool is pulled from the hole.

Figure 118. Bottomhole pressure record of a drill stem test

Coring

In addition to these tests, formation core samples are sometimes taken. Two methods of obtaining cores are frequently used. In one, an assembly called a *core barrel* is made up on the drill string and run to the bottom of the hole. As the core barrel is rotated, it cuts a cylindrical core a few inches in diameter that is received in a tube above the core-cutting bit. A complete round trip is required for each core taken. The second is a sidewall sampler in which a small explosive charge is fired to ram a small cylinder into the wall of the hole. When the tool is pulled out of the hole, the small core samples come out with the tool. Up to thirty of the small samples can be taken at any desired depth. Either type of core can be examined in a laboratory and may reveal much about the nature of the reservoir.

COMPLETING THE WELL

After the operating company carefully considers all the data obtained from the various tests it has ordered to be run on the formation or formations of interest, a decision is made on whether to set production casing and complete the well or to plug and abandon it. If the decision is to abandon it, the hole is considered to be *dry*, that is, not capable of producing oil or gas in commercial quantities. In other words, some oil or gas may be present but not in amounts great enough to justify the expense of completing the well. Therefore, several cement plugs will be set in the well to seal it off more or less permanently. However, sometimes wells that were plugged and abandoned as dry at one time in the past may be reopened and produced if the price of oil or gas has become more favorable. The cost of plugging and abandoning a well may only be a few thousand dollars. Contrast that cost with the price of setting a production string of casing—$50,000 or more. Therefore, the operator's decision is not always easy.

Setting Production Casing

If the operating company decides to set casing, casing will be brought to the well and for one final time, the casing and cementing crew run and cement a string of casing. Usually, the production casing is set and cemented through the pay zone; that is, the hole is drilled to a depth beyond the producing formation, and the casing is set to a point near the bottom of the hole. As a result, the casing and cement actually seal off the producing zone—but only temporarily. After the production string is cemented, the drilling contractor has almost finished his job except for a few final touches.

Perforating

Since the pay zone is sealed off by the production string and cement, perforations must be made in order for the oil or gas to flow into the wellbore. Perforations are simply holes that are made through the casing and cement and extend some distance into the formation. The most common method of perforating incorporates shaped-charge explosives (similar to those used in armor-piercing shells) (fig. 119).

Shaped charges accomplish penetration by creating a jet of high-pressure, high-velocity gas. The charges are arranged in a tool called a *gun* that is lowered into the well opposite the producing zone (fig. 120). Usually the gun is lowered in on wireline. When the gun is in position, the charges are fired by electronic means from the surface (fig. 121). After the perforations are made, the tool is retrieved. Perforating is usually performed by a service company that specializes in this technique.

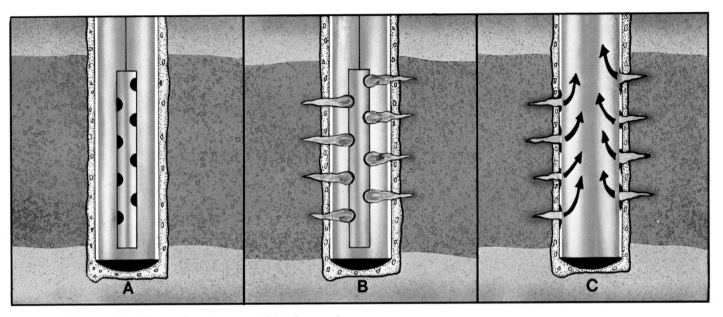

Figure 119. The perforating gun (A) is lowered into the wellbore to a depth opposite the producing zone. The shaped charges are fired (B), resulting in several perforations (C) that allow reservoir fluids to flow into the wellbore.

Figure 121. Perforating specialist actuates the switch to fire charges in a perforating gun.

Figure 120. Perforating guns are available in several sizes.

Figure 122. A Christmas tree of control valves is positioned on the completed well.

Installing the Christmas Tree

Even though the oil or gas can flow into the casing after it is perforated, usually, the well is not produced through the casing. Instead, small-diameter pipe called *tubing* is placed in the well to serve as a way for the oil or gas to flow to the surface. The tubing is run into the well with a packer. The packer goes on the outside of the tubing and is placed at a depth just above the producing zone. When the packer is expanded, it grips the wall of the production casing and forms a seal in the annular space between the outside of the tubing and the inside of the casing. Thus, as the produced fluids flow out of the formation through the perforations, they are forced to enter the tubing to get to the surface.

When casing is set, cemented, and perforated and when the tubing string is run, then a collection of valves called a *Christmas tree* is installed on the surface at the top of the casing (fig. 122). Like so many terms in the oil industry, no one knows why this device on top of the well is called a Christmas tree. Perhaps all the valves and piping reminded someone of a traditional Christmas tree. The tubing in the well is suspended from the Christmas tree, so as the well's production flows up the tubing, it enters the Christmas tree. As a result, the production from the well can be controlled by opening or closing valves on the Christmas tree.

Usually, once the Christmas tree is installed, the well can truly be said to be complete. The drilling contractor has done his job as called for in the drilling contract, and he can move the rig to another location to start the well-drilling process all over again.

Acidizing

Sometimes, however, petroleum exists in a formation but is unable to flow readily into the well because the formation has very low permeability. If the formation is composed of rocks that dissolve upon being contacted by acid, such as limestone or dolomite, then a technique known as acidizing may be required (fig. 123). Acidizing is usually performed by an acidizing service company and may be done before the rig is moved off the well; or it can also be done after the rig is moved away. In any case, the acidizing operation basically consists of pumping anywhere from fifty to thousands of gallons of

Figure 123. Acidizing a deep well

acid down the well. The acid travels down the tubing, enters the perforations, and contacts the formation. Continued pumping forces the acid into the formation where it etches channels—channels that provide a way for the formation's oil or gas to enter the well through the perforations (fig. 124).

Fracturing

When a formation contains oil or gas in commercial quantities but the permeability is too low to permit good recovery, a process called *fracturing* may be used to increase permeability to a practical level.
Basically, to fracture a formation, a fracturing service company pumps a specially blended fluid down the well and into the formation under great pressure. Pumping continues until the formation literally cracks open (fig. 125).

Meanwhile, sand, walnut hulls, or aluminum pellets are mixed into the fracturing fluid. These materials are called *proppants*. The proppant enters the fractures in the formation, and, when pumping is stopped and the pressure allowed to dissipate, the proppant remains in the fractures. Since the fractures try to close back together after the pressure on the well is released, the proppant is needed to hold or prop the fractures open. These propped-open fractures provide passages for oil or gas to flow into the well.

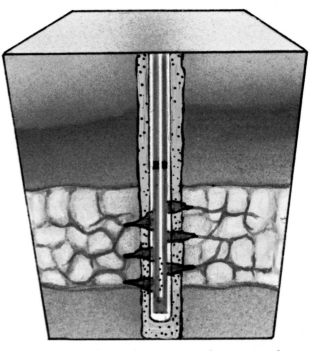

Figure 124. In acidizing, several pump trucks send acid down the well. The acid contacts the formation and etches channels into it. The channels provide a path for fluids to flow out of the formation and into the well.

Figure 125. In a fracturing job, several pump trucks force fluid containing a proppant down the well and into the perforations in the casing. Continued pumping builds up great pressure that causes the producing formation to fracture. When the pressure is allowed to dissipate, the proppant holds the fracture open and forms a permeable channel for formation fluids to flow into the wellbore.

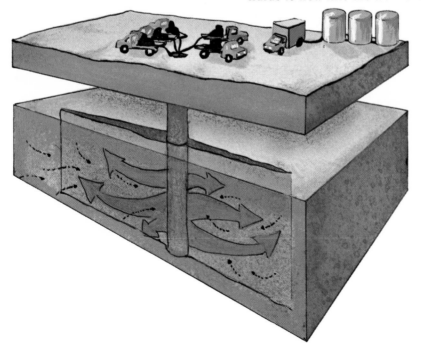

71

SPECIAL DRILLING OPERATIONS

Air Drilling

Sometimes, it is possible for the drilling contractor to circulate air instead of drilling mud (fig. 126). Sometimes formation conditions are such that mud is not really needed. There must be no danger of high-pressure formations being encountered, and formations containing a lot of water cannot be drilled with air. The trouble with too much water is that it mixes with the very fine, dustlike cuttings that are made when drilling with air. So, the cuttings ball up and can't get out of the hole. Drilling is impeded, and the drill string could even get stuck in the hole.

However, when it's possible to use air, drilling rates are faster than when mud is used. Mud, being heavier than air, has a tendency to hold down the cuttings made by the bit, so the bit spends a lot of time redrilling old cuttings as well as drilling new formation. When air is used, bit cuttings are not held on bottom but are immediately blown away. To drill with air, large compressors are moved onto the site, and the air is circulated down the drill string, out of the bit, and up the annulus as usual. Of course, there's no need to recirculate plain old air, so it and the cuttings carried up are blasted out through a *blooey line,* which is just a piece of pipe run out to the reserve pits (fig. 127).

Figure 126. This rig is equipped to drill holes using air as a circulating fluid. Note the air compressors at right.

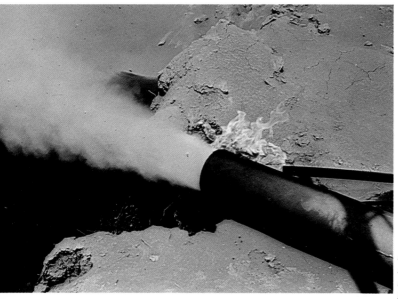

Figure 127. Cuttings are blasted out the blooey line. The flame is there as a precaution. Should gas be encountered, the flare will ignite it so that the gas will burn away harmlessly.

72

It's not often feasible to drill an entire hole using nothing but air. Usually, only a part of the hole is drilled with air, and, when it becomes necessary, the crew will *mud up,* that is, switch over to drilling mud. However, when it is possible, drilling with air certainly makes for fast drilling rates.

Directional Drilling

Usually, the crew tries to drill the hole as straight as possible. However, at times it's desirable to deflect the hole from vertical and drill it on a slant. Perhaps the most dramatic example of slant or directional drilling is on offshore drilling platforms. There a platform is erected over the drilling site, and several wells are drilled from this single platform without having to move it (fig. 128). The technique used is directional drilling.

Only the first hole drilled into the reservoir may be vertical; every subsequent well may be drilled vertically to a certain depth, then kicked off (deflected) directionally so that the bottom of the hole ends up perhaps hundreds of feet away from its starting point on the surface. By using directional drilling, as many as twenty or more wells may be drilled into the reservoir from one platform.

Directional drilling involves the use of some rather interesting downhole tools and techniques. For example, some means of kicking the hole off vertical must be used.

Figure 128. Several directional wells are often drilled from a single offshore platform.

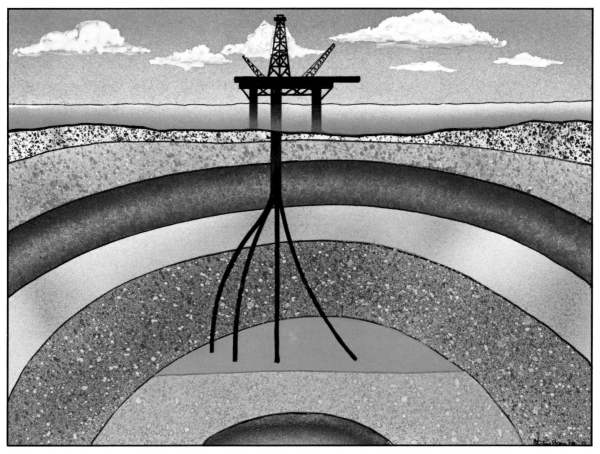

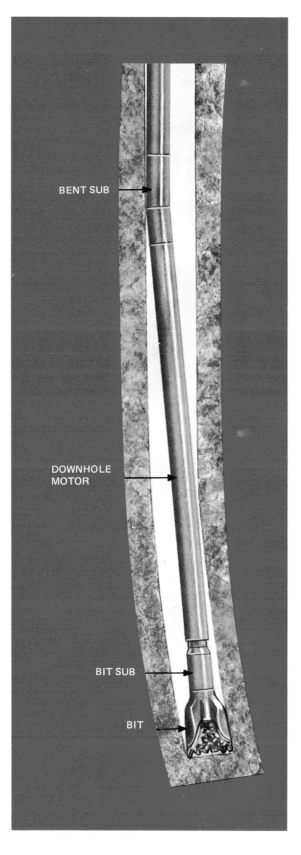

BENT SUB

DOWNHOLE
MOTOR

BIT SUB

BIT

This might be accomplished with a *bent sub* and a *downhole motor* (figs. 129 and 130). A sub (short for substitute) is a special device that is threaded so that it can be attached to or made up in the drill string. A bent sub is simply a short piece of pipe threaded on both ends, that has a bend in the middle. The bend has an angle of from 1 to 3 degrees. A downhole motor is a tool shaped like a piece of pipe that has turbine blades (a turbine is like a series of electric fan blades stacked on top of each other on a shaft), or it can be a multicurved steel shaft that turns inside an elliptically shaped opening in a housing. In practice, the bit is made up in the bottom of the downhole motor and the bent sub on the top. This assembly is tripped into the hole as usual.

When the tool reaches bottom, it must be oriented (pointed in the direction necessary to get the hole to go in the desired direction). To orient the tool, various types of compasses or directional gyroscopes, coupled with photographic or electronic readout devices, are employed. Once the tool is oriented, drilling begins. However, the drill string is not rotated. Instead, drilling mud flowing through the directional motor causes the turbine blades to turn, or the multicurved shaft to turn, which causes the bit to rotate. Because of the bent sub, the hole starts off at an angle, a relatively small angle (1 to 3 degrees) at first, but the angle is increased as drilling progresses—up to almost 90 degrees from vertical if necessary. Periodically, the hole is surveyed; that is, using the compass or an electronic readout device, its direction and angle of deflection is checked. The angle and direction of the hole are carefully maintained until total depth is reached and the pay zone is penetrated.

Another directional drilling tool is the old, reliable *whipstock*. A long, heavy steel member, the whipstock has an inclined or slanted face, much like a slanted ramp that people confined to wheelchairs use instead of stairs. The slanted face forces the bit to start drilling at an angle.

Figure 129. Positive-displacement motor deflected with bent sub

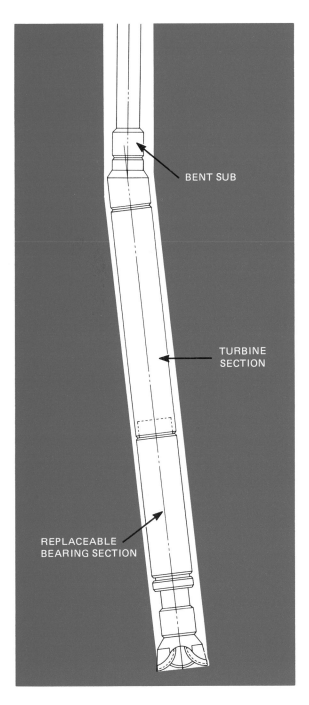

Figure 130. A turbine motor deflected with bent sub

Fishing

A *fish* is a piece of equipment, a tool, or a part or all of the drill string lost or stuck in the hole. Small pieces, such as a bit cone, a wrench, or any other relatively small, nondrillable item, are called *junk*. In any case, whenever there is junk or a fish in the hole, it must be removed or fished out so that drilling operations can continue.

Once again, a number of ingenious tools and techniques have been developed to retrieve a fish. For example, an *overshot* (fig. 131) can be run into the hole down to the fish.

Figure 131. An overshot is used to retrieve fish from the hole.

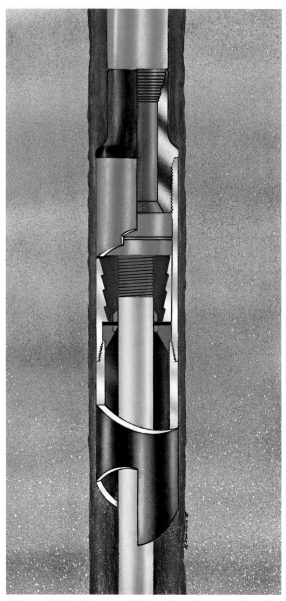

Figure 132. An overshot grips the outside of a fish for removal from the hole.

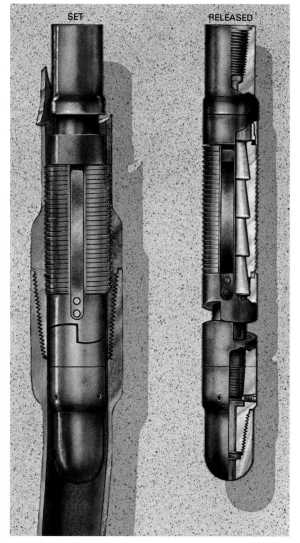

SET   RELEASED

Figure 133. A spear grips the inside of the fish.

and set around the outside of it. Grapples in the overshot latch onto the fish firmly and then the overshot and attached fish are pulled out of the hole (fig. 132).

Another fishing tool is a *spear* (fig. 133). Unlike an overshot, which goes over the outside of the fish, a spear goes inside. It grips the inside of the fish and allows it to be retrieved. Other fishing tools include powerful *magnets* and *baskets* through which mud can be circulated, both of which are useful for retrieving junk from the hole.

Special *mills* (fig. 134) are available that allow the jagged top of a fish to be ground smooth so that an overshot or spear can be better attached. Since two fishing jobs are seldom alike, various other fishing tools have been designed to meet the unique requirements of the rig fisherman.

Figure 134. Mills such as these are used to grind the jagged tops of fish to facilitate overshot or spear attachment.

THE FUTURE

If Colonel Drake, Uncle Billy Smith, Anthony Lucas, and the untold thousands of other drilling pioneers were alive today, they would no doubt be amazed by the proliferation of drilling tools and techniques that have been developed since those primitive days in Pennsylvania and Texas.

They would probably also be impressed by the boom and bust nature of the petroleum industry. At one point in the early 1980s, over 4,500 drilling rigs were at work in the United States. By the mid-1980s, the number had shrunk to a record low—below 700. Yet the survivors of the bust remained optimistic about the future; just as there had been busts, they reasoned, there had also been booms. They believed that the industry would come back—it always had before, and it would again.

The logic behind such optimism was solid, for they considered the following points. First, oil and gas exist in the earth in limited amounts; hydrocarbons are a finite energy source. Second, oil and gas are the preferred energy sources of the U.S. and the world. Third, as the price of oil goes down, consumption goes up. In the 1960s and 1970s, when oil prices hovered around $2.50 to $3.00 per barrel (1 barrel equals 42 gallons), consumption rose at a rapid rate, so that even with an Arab oil embargo that drove prices up to a high of about $35.00 per barrel, consumption in the U.S. peaked at over 18 million barrels per day by the late 1970s. As the price of oil remained high, consumption went down, so that by the mid-1980s, one day's supply for the U.S. was about 16 million barrels. With less oil being consumed, and with many of the reservoirs that were discovered during the drilling boom coming on line with production, an oversupply of oil resulted, and prices went down. With

prices down, consumption started back up; thus, there was every reason to believe that in the not-too-distant future, shortages would again occur, prices would go higher, and drilling activity would pick up.

A fourth consideration for the likelihood of a return to more drilling activity was the fact that the U.S. used more oil than it produced from domestic wells; that is, the country used so much oil that it had to import, on the average, about one-third to one-half of the amount it consumed.

Depending on imported oil made many people in the oil industry uneasy. They asked, "How can we continue to rely on foreign sources for much of our oil supply? If, as happened in 1973, a foreign government in control of its oil production can create an oil shortage, should not we here in the U.S. concentrate our efforts on finding new supplies on our own ground?" While members of the oil industry answered the question with a resounding *yes,* the general public and members of the U.S. Congress were not so sure. Oil was cheap and plentiful. The industry believed, however, that when oil prices went up—as they were bound to when the supply fell short—domestic drilling would be encouraged.

While it is interesting to speculate on what will happen to the oil business in the future, little doubt exists that with present technology, oil and gas remain the stalwart forms of energy. Until viable alternative energy sources can be found and implemented, petroleum and petroleum derivatives are likely to be the leading sources. Therefore, the sight of a drilling rig in the distance, starkly silhouetted against a lonely sky, is likely to remain a familiar scene for many years to come.

GLOSSARY

A

abandon *v:* to cease producing oil and gas from a well when it becomes unprofitable. A wildcat well may be abandoned after it has proven nonproductive. Several steps are involved in abandoning a well: part of the casing is removed and salvaged; one or more cement plugs are placed in the borehole to prevent migration of fluids between the different formations penetrated by the borehole; and the well is abandoned. In many states, it is necessary to secure permission from official agencies before a well may be abandoned.

absolute permeability *n:* a measure of the ability of a single fluid (as water, gas, or oil) to flow through a rock formation when the formation is totally filled (saturated) with the single fluid. The permeability measure of a rock filled with a single fluid is different from the permeability measure of the same rock filled with two or more fluids. Compare *effective permeability.*

acid fracture *v:* to part or open fractures in productive, hard-limestone formations by using a combination of oil and acid or water and acid under high pressure. See *formation fracturing.*

acidize *v:* to treat oil-bearing limestone or other formations, using a chemical reaction with acid, to increase production. Hydrochloric or other acid is injected into the formation under pressure. The acid etches the rock, enlarging the pore spaces and passages through which the reservoir fluids flow. The acid is held under pressure for a period of time and then pumped out, and the well is swabbed and put back into production. Chemical inhibitors combined with the acid prevent corrosion of the pipe.

adjustable choke *n:* a choke in which a conical needle and seat vary the rate of flow. See *choke.*

air-actuated *adj:* powered by compressed air, as the clutch and brake system in drilling equipment.

air drilling *n:* a method of rotary drilling that uses compressed air as the circulation medium. The conventional method of removing cuttings from the wellbore is to use a flow of water or drilling mud. Compressed air removes the cuttings with equal or greater efficiency. The rate of penetration is usually increased considerably when air drilling is used. However, a principal problem in air drilling is the penetration of formations containing water, since the entry of water into the system reduces the ability of the air to remove the cuttings.

American Petroleum Institute *n:* 1. founded in 1920, this national oil trade organization is the leading standardizing organization on oil field drilling and producing equipment. It maintains departments of transportation, refining, and marketing in Washington, D.C., and a department of production in Dallas. 2. (slang) indicative of a job being properly or thoroughly done (as, "His work is strictly API"). 3. degrees API; used to designate API gravity. See *API gravity.*

angle of deflection *n:* in directional drilling, the angle, expressed in degrees, at which a well is deflected from the vertical by a whipstock or other deflecting tool. See *whipstock.*

annular blowout preventer *n:* a large valve, usually installed above the ram preventers, that forms a seal in the annular space between the pipe and wellbore or, if no pipe is present, on the wellbore itself. Compare *ram blowout preventer.*

annular space *n:* 1. the space surrounding a cylindrical object within a cylinder. 2. the space around a pipe in a wellbore, the outer wall of which may be the wall of either the borehole or the casing; sometimes termed the annulus.

anticline *n:* an arched, inverted-trough configuration of folded and stratified rock layers. Compare *syncline.*

API *abbr:* American Petroleum Institute.

API gravity *n:* the measure of the density or gravity of liquid petroleum products in the United States, derived from specific gravity in accordance with the following equation:

$$\text{API gravity} = \frac{141.5}{\text{specific gravity}} - 131.5.$$

API gravity is expressed in degrees, a specific gravity of 1.0 being equivalent to 10° API.

B

back off *v:* to unscrew one threaded piece (as a section of pipe) from another.

back up *v:* to hold one section of an object (as pipe) while another is being screwed into or out of it.

bail *n:* a cylindrical steel bar (similar to the handle or bail of a bucket, only much larger) that supports the swivel and connects it to the hook. Sometimes, the two cylindrical bars that support the elevators and attach them to the hook are called bails. *v:* to recover bottomhole fluids, samples, or drill cuttings by lowering a cylindrical vessel called a bailer to the bottom of a well, filling it, and retrieving it. See *bailer*.

bailer *n:* a long cylindrical container, fitted with a valve at its lower end, used to remove water, sand, mud, or oil from a well.

bailing line *n:* cable attached to the bailer, passed over a sheave at the top of the derrick, and spooled on a reel. See *sheave*.

barge *n:* any one of many types of flat-decked, shallow draft vessels, usually towed by a boat. A complete drilling rig may be assembled on a drilling barge, which usually is submersible; that is, it has a submersible hull or base that is flooded with water at the drilling site. Drilling equipment, crew quarters, and so forth are mounted on a superstructure above the water level.

barite or **baryte** *n:* barium sulfate, $BaSO_4$; a mineral used to increase the weight of drilling mud. Its specific gravity is 4.2 (i.e., it is 4.2 times heavier than water). See *barium sulfate* and *mud*.

barium sulfate *n:* 1. a chemical combination of barium, sulfur, and oxygen. Also called barite. See *barite*. 2. a tenacious scale that is very difficult to remove.

barrel *n:* a measure of volume for petroleum products. One barrel is the equivalent of 42 U.S. gallons or 0.15899 cubic metres. One cubic metre equals 6.2897 barrels.

basket sub *n:* a fishing accessory run above a bit or mill to recover small pieces of metal or junk in a well.

bed *n:* a specific layer of earth or rock in contrast to other layers of different material lying above, below, or adjacent to it.

belt *n:* a flexible band or cord connecting and passing about each of two or more pulleys to transmit power or impart motion.

bit *n:* the cutting or boring element used in drilling oil and gas wells. The bit consists of the cutting element and the circulating element. The circulating element permits the passage of drilling fluid and utilizes the hydraulic force of the fluid stream to improve drilling rates. In rotary drilling, several drill collars are joined to the bottom end of the drill-pipe column. The bit is attached to the end of the drill collar. Most bits used in rotary drilling are roller cone bits.

bit breaker *n:* a heavy plate that fits in the rotary table and holds the drill bit while it is being made up in or broken out of the drill stem. See *bit*.

bit record *n:* a report on each bit used in a drilling operation that lists the bit type, the amount of footage the bit has drilled, and the nature of the formation penetrated.

blind ram *n:* an integral part of a blowout preventer that serves as the closing element. Its ends do not fit around the drill pipe but seal against each other and shut off the space below completely.

block *n:* any assembly of pulleys on a common framework; in mechanics, one or more pulleys, or sheaves, mounted to rotate on a common axis. The crown block is an assembly of sheaves mounted on beams at the top of the derrick. The drilling line is reeved over the sheaves of the crown block alternately with the sheaves of the traveling block, which is hoisted and lowered in the derrick by the drilling line. When elevators are attached to a hook on the traveling block, and when drill pipe is latched in the elevators, the pipe can be raised or lowered in the derrick or mast. See *crown block, elevator, hook, reeve, sheave,* and *traveling block;* also see *drilling block*.

blooey line *n:* the discharge pipe from a well being drilled by air drilling. The blooey line is used to conduct the air or gas used for circulation away from the rig to reduce the fire hazard as well as to transport the cuttings a suitable distance from the well. See *air drilling*.

blowout *n:* an uncontrolled flow of gas, oil, or other well fluids into the atmosphere. A blowout, or gusher, occurs when formation pressure exceeds the pressure applied to it by the column of drilling fluid. A kick warns of an impending blowout. See *formation pressure, gusher,* and *kick*.

blowout preventer *n:* one of several valves installed at the wellhead to prevent the escape of pressure either in the annular space between the casing and drill pipe or in open hole (i.e., hole with no drill pipe) during drilling or completion operations. Blowout preventers on land rigs are located beneath the rig at the land's surface; on jackup or platform rigs, they are located at the water's surface; and on floating offshore rigs, on the seafloor. See *annular blowout preventer, inside blowout preventer,* and *ram blowout preventer*.

boll-weevil *n:* (slang) an inexperienced rig or oil-field worker, sometimes shortened to "weevil."

bomb *n:* a thick-walled container, usually steel, used to hold samples of oil or gas under pressure. See *bottom-hole pressure*.

bond *n:* the state of one material adhering or being joined to another material (as cement to formation). *v:* to adhere or be joined to another material.

BOP *abbr:* blowout preventer.

borehole *n:* the wellbore; the hole made by drilling or boring. See *wellbore*.

bottomhole *n:* the lowest or deepest part of a well. *adj:* pertaining to the bottom of the wellbore.

bottomhole choke *n:* a device with a restricted opening placed in the lower end of the tubing to control the rate of flow. See *choke*.

bottomhole pressure *n:* 1. the pressure at the bottom of a borehole. It is caused by the hydrostatic pressure of the drilling fluid in the hole and, sometimes, any back-pressure held at the surface as when the well is shut in with blowout preventers. When mud is being circulated, bottomhole pressure is the hydrostatic pressure plus the remaining circulating pressure required to move the mud up the annulus. 2. the pressure in a well at a point opposite the producing formation, as recorded by a bottomhole pressure bomb.

box *n:* the female section of a tool joint. See *tool joint*.

brake *n:* a device for arresting the motion of a mechanism, usually by means of friction, as in the drawworks brake. Compare *electrodynamic brake* and *hydromatic brake*.

break out *v:* 1. to unscrew one section of pipe from another section, especially drill pipe while it is being withdrawn from the wellbore. During this operation, the tongs are used to start the unscrewing operation. See *tongs*. 2. to separate, as gas from liquid.

breakout cathead *n:* a device attached to the shaft of the drawworks that is used as a power source for unscrewing drill pipe; usually located opposite the driller's side of the drawworks. See *cathead.*

breakout tongs *n:* tongs that are used to start unscrewing one section of pipe from another section, especially drill pipe coming out of the hole. Also called lead tongs. See *tongs.*

bring in a well *v:* to complete a well and put it in producing status.

buck up *v:* to tighten up a threaded connection (as two joints of drill pipe).

bullet perforator *n:* a tubular device that, when lowered to a selected depth within a well, fires bullets through the casing to provide holes through which the well fluids may enter.

C

cable *n:* a rope of wire, hemp, or other strong fibers. See *wire rope.*

cable-tool drilling *n:* a drilling method in which the hole is drilled by dropping a sharply pointed bit on the bottom of the hole. The bit is attached to a cable, and the cable is picked up and dropped, picked up and dropped, over and over, as the hole is drilled.

cap rock *n:* 1. impermeable rock overlying an oil or gas reservoir that tends to prevent migration of oil or gas out of the reservoir. 2. the porous and permeable strata overlying salt domes that may serve as the reservoir rock.

cased *adj:* pertaining to a wellbore in which casing is run and cemented. See *casing.*

cased hole *n:* a wellbore in which casing has been run.

casing *n:* steel pipe placed in an oil or gas well as drilling progresses to prevent the wall of the hole from caving in during drilling and to provide a means of extracting petroleum if the well is productive.

casing centralizer *n:* a device secured around the casing at regular intervals to center it in the hole. Casing that is centralized allows a more uniform cement sheath to form around the pipe.

casing coupling *n:* a tubular section of pipe that is threaded inside and used to connect two joints of casing.

casing elevator *n:* See *elevator.*

casinghead *n:* a heavy, steel, flanged fitting that connects to the first string of casing and provides a housing for the slips and packing assemblies by which intermediate strings of casing are suspended and the annulus sealed off. Also called a spool. See *annular space.*

casing shoe *n:* also called a guide shoe. See *guide shoe.*

casing string *n:* the entire length of all the joints of casing run in a well. Casing is manufactured in lengths of about 30 feet, each length or joint being joined to another as casing is run in a well. See *combination string.*

catch samples *v:* to obtain cuttings for geological information as formations are penetrated by the bit. The samples are obtained from drilling fluid as it emerges from the wellbore or, in cable-tool drilling, from the bailer. Cuttings are carefully washed until they are free of foreign matter, dried, and labeled to indicate the depth at which they were obtained. See *bailer, cable-tool drilling,* and *cuttings.*

cathead *n:* a spool-shaped attachment on a winch around which rope for hoisting and pulling is wound. See *breakout cathead* and *makeup cathead.*

catline *n:* a hoisting or pulling line powered by the cathead and used to lift heavy equipment on the rig. See *cathead.*

caving *n:* collapse of the walls of the wellbore, also called sloughing.

cellar *n:* a pit in the ground to provide additional height between the rig floor and the wellhead to accommodate the installation of blowout preventers, rathole, mousehole, and so forth. It also collects drainage water and other fluids for subsequent disposal.

cement casing *v:* to fill the annulus between the casing and hole with cement to support the casing and prevent fluid migration between permeable zones.

cement channeling *n:* an undesirable phenomenon that can occur when casing is being cemented in a borehole. The cement slurry fails to rise uniformly between the casing and borehole wall, leaving spaces void of cement. Ideally, the cement should completely and uniformly surround the casing and form a strong bond to the borehole wall.

cementing *n:* the application of a liquid slurry of cement and water to various points inside or outside the casing. See *primary cementing, secondary cementing,* and *squeeze cementing.*

chain drive *n:* a drive system using a chain and chain gears to transmit power. Power transmissions use a roller chain, in which each link is made of side bars, transverse pins, and rollers on the pins. A double roller chain is made of two connected rows of links, a triple roller chain of three, and so forth.

chain tongs *n:* a tool consisting of a handle and releasable chain used for turning pipe or fittings of a diameter larger than that which a pipe wrench would fit. The chain is looped and tightened around the pipe or fitting, and the handle is used to turn the tool so that the pipe or fitting can be tightened or loosened.

check valve *n:* a valve that permits flow in one direction only.

choke *n:* an orifice installed in a line to restrict the flow and control the rate of production. Surface chokes are part of the Christmas tree and contain a choke nipple, or bean, with a small-diameter bore that serves to restrict the flow. Chokes are also used to control the rate of flow of the drilling mud out of the hole when the well is closed in with the blowout preventer and a kick is being circulated out of the hole. See *adjustable choke, blowout preventer, bottomhole choke, Christmas tree, kick, nipple,* and *positive choke.*

choke line *n:* an extension of pipe from the blowout preventer assembly used to direct well fluids from the annulus to the choke manifold.

choke manifold *n:* the arrangement of piping and special valves, called chokes, through which drilling mud is circulated when the blowout preventers are closed to control the pressures encountered during a kick. See *choke* and *blowout preventer.*

Christmas tree *n:* the control valves, pressure gauges, and chokes assembled at the top of a well to control the flow of oil and gas after the well has been drilled and completed.

circulate *v:* to pass from one point throughout a system and back to the starting point. For example, drilling fluid is circulated out of the suction pit, down the drill pipe and drill collars, out the bit, up the annulus, and back to the pits.

circulation *n:* the movement of drilling fluid out of the mud pits, down the drill stem, up the annulus, and back to the mud pits.

combination string *n:* a casing string that has joints of various collapse resistance, internal yield strength, and tensile strength designed for various depths in a specific well to best withstand the conditions of that well. In deep wells, high tensile strength is required in the top casing joints to carry the load, whereas high collapse resistance and internal yield strength are needed for the bottom joints. In the middle of the casing, average qualities are usually sufficient. The most suitable combination of types and weights of pipe helps to ensure efficient production at a minimum cost.

come out of the hole *v:* to pull the drill stem out of the wellbore. This withdrawal is necessary to change the bit, change from a core barrel to the bit, run electric logs, prepare for a drill stem test, run casing, and so on.

company man *n:* also called company representative. See *company representative.*

company representative *n:* an employee of an operating company whose job is to represent the company's interests at the drilling location.

complete a well *v:* to finish work on a well and bring it to productive status. See *well completion.*

compound *n:* a mechanism used to transmit power from the engines to the pump, drawworks, and other machinery on a drilling rig. It is composed of clutches, chains and sprockets, belts and pulleys, and a number of shafts, both driven and driving. *v:* to connect two or more power-producing devices (as engines) to run one piece of driven equipment (as the drawworks).

conductor pipe *n:* 1. a short string of large-diameter casing used to keep the top of the wellbore open and to provide a means of conveying the up-flowing drilling fluid from the wellbore to the mud pit.

contract depth *n:* the depth of the wellbore at which the drilling contract is fulfilled.

core *n:* a cylindrical sample taken from a formation for geological analysis. Usually a conventional core barrel is substituted for the bit and procures a sample as it penetrates the formation. See also *sidewall coring. v:* to obtain a formation sample for analysis.

core analysis *n:* laboratory analysis of a core sample to determine porosity, permeability, lithology, fluid content, angle of dip, geological age, and probable productivity of the formation.

core barrel *n:* a tubular device from 25 to 60 feet long run at the bottom of the drill pipe in place of a bit to cut a core sample.

core catcher *n:* the part of the core barrel that holds the formation sample.

core cutterhead *n:* the cutting element of the core barrel assembly. In design it corresponds to one of the three main types of bits: drag bits with blades for cutting soft formations; roller bits with rotating cutters for cutting medium hard formations; and diamond bits for cutting very hard formations.

coupling *n:* 1. in piping, a metal collar with internal threads used to join two sections of threaded pipe. 2. in power transmission, a connection extending longitudinally between a driving shaft and a driven shaft. Most such couplings are flexible and compensate for minor misalignment of the two shafts.

crooked hole *n:* a wellbore that has deviated from the vertical. It usually occurs in areas where the subsurface formations are difficult to drill, such as a section of alternating hard and soft strata steeply inclined from the horizontal.

crown block *n:* an assembly of sheaves or pulleys mounted on beams at the top of the derrick over which the drilling line is reeved. See *block, reeve,* and *sheave.*

cuttings *n pl:* the fragments of rock dislodged by the bit and brought to the surface in the drilling mud. Washed and dried samples of the cuttings are analyzed by geologists to obtain information about the formations drilled.

D

daylight tour *n:* (pronounced "tower") the shift of duty on a drilling rig that starts at or about daylight; also called morning tour. Compare *evening tour* and *graveyard tour.*

deadline *n:* the drilling line from the crown block sheave to the anchor, so called because it does not move. Compare *fast line.*

deadline tie-down anchor *n:* a device to which the deadline is attached, securely fastened to the mast or derrick substructure. Also called a deadline anchor.

degasser *n:* the equipment used to remove unwanted gas from a liquid, especially from drilling fluid.

density *n:* the mass or weight of a substance; often expressed in weight per unit volume. For instance, the density of a drilling mud may be 10 pounds per gallon (ppg), 74.8 pounds per cubic foot (lb/ft³), or 1,198.2 kilograms per cubic metre (kg/m³). Specific gravity and API gravity are other units of density. See *API gravity* and *specific gravity.*

derrick *n:* a large load-bearing structure, usually of bolted construction. In drilling, the standard derrick has four legs standing at the corners of the substructure and reaching to the crown block. The substructure is an assembly of heavy beams used to elevate the derrick and provide space to install blowout preventers, casingheads, and so forth. Because the standard derrick must be assembled piece by piece, it has largely been replaced by the mast, which can be lowered and raised without disassembly. See *crown block, mast,* and *substructure.*

derrickman *n:* the crew member who handles the upper end of the drill stem as it is being hoisted out of or lowered into the hole. He is also responsible for the conditioning of the drilling fluid and the circulating machinery.

desander *n:* a centrifugal device for removing sand from drilling fluid to prevent abrasion of the pumps. It may be operated mechanically or by a fast-moving stream of fluid inside a special cone-shaped vessel, in which case it is sometimes called a hydrocyclone. Compare *desilter.*

desilter *n:* a centrifugal device for removing very fine particles, or silt, from drilling fluid to keep the amount of solids in the fluid to the lowest possible point. Usually, the lower the solids content of mud, the faster the rate of penetration. It works on the same principle as a desander. Compare *desander.*

development well *n:* 1. a well drilled in proven territory in a field to complete a pattern of production. 2. an exploitation well. See *exploitation well.*

deviation *n:* the inclination of the wellbore from the vertical. The angle of deviation, angle of drift, or drift angle is the angle in degrees that shows the variation from the vertical as revealed by a deviation survey. See *deviation survey.*

deviation survey *n:* an operation made to determine the angle from which a bit has deviated from the vertical during drilling. There are two basic deviation survey, or drift survey, instruments: one reveals the angle of deviation only; the other indicates both the angle and direction of deviation.

diamond bit *n:* a drilling bit that has a steel body surfaced with industrial diamonds. Cutting is performed by the rotation of the very hard diamonds over the rock surface.

diesel-electric power *n:* the power supplied to a drilling rig by diesel engines driving electric generators, used widely offshore and gaining popularity onshore.

diesel engine *n:* a high-compression, internal-combustion engine used extensively for powering drilling rigs. In a diesel engine, air is drawn into the cylinders and compressed to very high pressures; ignition occurs as fuel is injected into the compressed and heated air. Combustion takes place within the cylinder above the piston, and expansion of the combustion products imparts power to the piston.

directional drilling *n:* intentional deviation of a wellbore from the vertical. Although wellbores are normally drilled vertically, it is sometimes necessary or advantageous to drill at an angle from the vertical. Controlled directional drilling makes it possible to reach subsurface areas laterally remote from the point where the bit enters the earth. It involves the use of turbodrills, Dyna-Drills, whipstocks, or other deflecting tools. See *Dyna-Drill, turbodrill,* and *whipstock.*

discovery well *n:* the first oil or gas well drilled in a new field; the well that reveals the presence of a petroleum-bearing reservoir. Subsequent wells are development wells. Compare *development well.*

displacement fluid *n:* in oilwell cementing, the fluid, usually drilling mud or salt water, that is pumped into the well after the cement to force the cement out of the casing and into the annulus.

doghouse *n:* 1. a small enclosure on the rig floor used as an office for the driller or as a storehouse for small objects. 2. any small building used as an office or for storage.

double *n:* a length of drill pipe, casing, or tubing consisting of two joints screwed together. Compare *thribble* and *fourble.* See *joint.*

double board *n:* the name used for working platform of the derrickman, or monkeyboard, when it is located at a height in the derrick or mast equal to two lengths of pipe joined together. Compare *fourble board* and *thribble board.* See *monkeyboard.*

drawworks *n:* the hoisting mechanism on a drilling rig. It is essentially a large winch that spools off or takes in the drilling line and thus raises or lowers the drill stem and bit.

drill bit *n:* the cutting or boring element used for drilling. See *bit.*

drill collar *n:* a heavy, thick-walled tube, usually steel, used between the drill pipe and the bit in the drill stem. Drill collars are used to put weight on the bit so that the bit can drill.

driller *n:* the employee directly in charge of a drilling rig and crew. His main duty is operation of the drilling and hoisting equipment, but he is also responsible for the downhole condition of the well, operation of downhole tools, and pipe measurements.

drilling block *n:* a lease or a number of leases of adjoining tracts of land that constitute a unit of acreage sufficient to justify the expense of drilling a wildcat.

drilling contractor *n:* an individual or group of individuals that own a drilling rig or rigs and contract their services for drilling wells to a certain depth.

drilling crew *n:* a driller, a derrickman, and two or more helpers who operate a drilling rig for one tour each day. See *derrickman, driller,* and *tour.*

drilling fluid *n:* circulating fluid, one function of which is to force cuttings out of the wellbore and to the surface. While a mixture of clay, water, and other chemical additives is the most common drilling fluid, wells can also be drilled using air, gas, or water as the drilling fluid. Also called circulating fluid. See *mud.*

drilling foreman *n:* the supervisor of drilling operations on a rig; also the tool pusher or rig superintendent.

drilling line *n:* a wire rope used to support the drilling tools.

drilling rate *n:* the speed with which the bit drills the formation; usually called the rate of penetration.

drilling rig *n:* See *rig.*

drill pipe *n:* the heavy seamless tubing used to rotate the bit and circulate the drilling fluid. Joints of pipe 30 feet long are coupled together by means of tool joints.

drill ship *n:* a ship constructed to permit a well to be drilled from it at an offshore location. While not as stable as other floating structures (as a semisubmersible), drill ships, or shipshapes, are capable of drilling exploratory wells in relatively deep waters. They may have a ship hull, a catamaran hull, or a trimaran hull. See *semisubmersible drilling rig.*

drill stem *n:* all members in the assembly used for drilling by the rotary method from the swivel to the bit, including the kelly, drill pipe and tool joints, drill collars, stabilizers, and various subsequent items. Compare *drill string.*

drill-stem test *n:* a method of gathering data on the potential productivity of a formation before installing casing in a well. See *formation testing.*

drill string *n:* the column, or string, of drill pipe with attached tool joints that transmits fluid and rotational power from the kelly to the drill collars and bit. Often, especially in the oil patch, the term is loosely applied to include both drill pipe and drill collars. Compare *drill stem.*

drum *n:* 1. a cylinder around which wire rope is wound in the drawworks. The drawworks drum is that part of the hoist upon which the drilling line is wound. 2. a steel container of general cylindrical form. Refined products are shipped in steel drums with capacities of about 50 to 55 U.S. gallons (about 200 litres).

DST *abbr:* drill-stem test.

Dyna-Drill *n:* a downhole motor driven by drilling fluid that imparts rotary motion to a drilling bit connected to the tool, thus eliminating the need to turn the entire drill stem to make hole. The Dyna-Drill, a trade name, is used in straight and directional drilling.

dynamic positioning *n:* a method by which a floating offshore drilling rig is maintained in position over an offshore well location. Generally, several motors called thrusters are located on the hull(s) of the structure and are actuated by a sensing system. A computer to which the system feeds signals then directs the thrusters to maintain the rig on location.

E

effective permeability *n:* a measure of the ability of a single fluid to flow through a rock when the pore spaces of the rock are not completely filled or saturated with the fluid. Compare *absolute permeability* and *relative permeability.*

electric well log *n:* a record of certain electrical characteristics of formations traversed by the borehole, made to identify the formations, determine the nature and amount of fluids they contain, and estimate their depth. Also called an electric log or electric survey.

electrodynamic brake *n:* a device mounted on the end of the drawworks shaft of a drilling rig. The electrodynamic brake (sometimes called a magnetic brake) serves as an auxiliary to the mechanical brake when pipe is lowered into a well. The braking effect in an electrodynamic brake is achieved by means of the interaction of electric currents with magnets, with other currents, or with themselves.

elevator *n:* a set of clamps that grips a stand, or column, of casing, tubing, or drill pipe so that the stand can be raised or lowered into the hole.

evening tour *n:* (pronounced "tower") the shift of duty on a drilling rig that starts in the afternoon and runs through the evening. Compare *daylight tour* and *graveyard tour*.

exploitation well *n:* a well drilled to permit more effective extraction of oil from a reservoir. It is sometimes called a development well. See *development well*.

exploration well *n:* a wildcat. See *wildcat*.

F

fastline *n:* the end of the drilling line that is affixed to the drum or reel of the drawworks, so called because it travels with greater velocity than any other portion of the line. Compare *deadline*.

fault *n:* a break in subsurface strata. Often strata on one side of the fault line have been displaced (upward, downward, or laterally) relative to their original positions.

field *n:* a geographical area in which a number of oil or gas wells produce from a continuous reservoir. A field may refer to surface area only or to underground productive formations as well. In a single field, there may be several separate reservoirs at varying depths.

fill the hole *v:* to pump drilling fluid into the wellbore while the pipe is being withdrawn in order to ensure that the wellbore remains full of fluid even though the pipe is withdrawn. Filling the hole lessens the danger of blowout or of caving of the wall of the wellbore.

filter cake *n:* 1. compacted solid or semisolid material remaining on a filter after pressure filtration of mud with the standard filter press. Thickness of the cake is reported in thirty-seconds of an inch or in millimetres. 2. the layer of concentrated solids from the drilling mud that forms on the walls of the borehole opposite permeable formations; also called wall cake or mud cake.

fingerboard *n:* a rack that supports the tops of the stands of pipe being stacked in the derrick or mast. It has several steel fingerlike projections that form a series of slots into which the derrickman can set a stand of drill pipe as it is pulled out of the hole.

fish *n:* an object left in the wellbore during drilling operations that must be recovered or drilled around before work can proceed. It can be anything from a piece of scrap metal to a part of the drill stem. *v:* 1. to recover from a well any equipment left there during drilling operations, such as a lost bit or drill collar or part of the drill string. 2. to remove from an older well certain pieces of equipment, such as packers, liners, or screen pipe, to allow reconditioning of the well.

fishing tool *n:* a tool designed to recover equipment lost in the well.

float collar *n:* a special coupling device, inserted one or two joints above the bottom of the casing string, that contains a check valve to permit fluid to pass downward but not upward through the casing. The float collar prevents drilling mud from entering the casing while it is being lowered, allowing the casing to float during its descent, which decreases the load on the derrick. The float collar also prevents a backflow of cement during the cementing operation.

floorman *n:* a drilling crew member whose work station is on the derrick floor. On rotary drilling rigs, there are at least two and usually three or more floormen on each crew. Also called rotary helper and roughneck.

fluid *n:* a substance that flows and yields to any force tending to change its shape. Liquids and gases are fluids.

formation *n:* a bed or deposit composed throughout of substantially the same kinds of rock; a lithologic unit. Each different formation is given a name, frequently as a result of the study of the formation outcrop at the surface and sometimes based on fossils found in the formation.

formation fracturing *n:* a method of stimulating production by increasing the permeability of the producing formation. Under extremely high hydraulic pressure, a fluid (as water, oil, alcohol, dilute hydrochloric acid, liquefied petroleum gas, or foam) is pumped downward through tubing or drill pipe and forced into the perforations in the casing. The fluid enters the formation and parts or fractures it. Sand grains, aluminum pellets, glass beads, or similar materials are carried in suspension by the fluid into the fractures. These are called propping agents or proppants. When the pressure is released at the surface, the fracturing fluid returns to the well, and the fractures partially close on the proppants, leaving channels for oil to flow through them to the well. This process is often called a frac job. See *propping agent*.

formation pressure *n:* the force exerted by fluids in a formation, recorded in the hole at the level of the formation with the well shut in. It is also called reservoir pressure or shut-in bottom-hole pressure. See *reservoir pressure* and *shut-in bottom-hole pressure*.

formation testing *n:* the gathering of data on a formation to determine its potential productivity before installing casing in a well. The conventional method is the drill stem test. Incorporated in the drill stem testing tool are a packer, valves or ports that may be opened and closed from the surface, and a pressure-recording device. The tool is lowered to bottom on a string of drill pipe and the packer set, isolating the formation to be tested from the formations above and supporting the fluid column above the packer. A port on the tool is opened to allow the trapped pressure below the packer to bleed off into the drill pipe, gradually exposing the formation to atmospheric pressure and allowing the well to produce to the surface, where the well fluids may be sampled and inspected. From a record of the pressure readings, a number of facts about the formation may be inferred.

fourble *n:* a section of drill pipe, casing, or tubing consisting of four joints screwed together. Compare *double* and *thribble*. See *joint*.

fourble board *n:* the name used for the working platform of the derrickman, or monkeyboard, when it is located at a height in the derrick equal to approximately four lengths of pipe joined together. Compare *double board* and *thribble board*. See *monkeyboard*.

fracturing *n:* shortened form of formation fracturing. See *formation fracturing*.

G

gas-cut mud *n:* a drilling mud that has entrained formation gas giving the mud a characteristically fluffy texture. When entrained gas is not released before the fluid returns to the well, the weight or density of the fluid column is reduced. Because a large amount of gas in mud lowers its density, gas-cut mud must be treated to lessen the chance of a blowout.

gas sand *n:* a stratum of sand or porous sandstone from which natural gas is obtained.

gas show *n:* the gas that appears in drilling fluid returns, indicating the presence of a gas zone.

geologist *n:* a scientist who gathers and interprets data pertaining to the strata of the earth's crust.

geology *n:* the science that relates to the study of the structure, origin, history, and development of the earth and its inhabitants as revealed in the study of rocks, formations, and fossils.

graveyard tour *n:* (pronounced "tower") the shift of duty on a drilling rig that starts at or about midnight. Compare *daylight tour* and *evening tour.*

gravity *n:* the attraction exerted by the earth's mass on objects at its surface; the weight of a body. See *API gravity* and *specific gravity.*

guide shoe *n:* a short, heavy, cylindrical section of steel filled with concrete and rounded at the bottom, which is placed at the end of the casing string. It prevents the casing from snagging on irregularities in the borehole as it is lowered. A passage through the center of the shoe allows drilling fluid to pass up into the casing while it is being lowered and cement to pass out during cementing operations. Also called casing shoe.

gun-perforate *v:* to create holes in casing and cement set through a productive formation. A common method of completing a well is to set casing through the oil-bearing formation and cement it. A perforating gun is then lowered into the hole and fired to detonate high-powered jets or shoot steel projectiles (bullets) through the casing and cement and into the pay zone. The formation fluids flow out of the reservoir through the performations and into the wellbore. See *jet-perforate* and *perforating gun.*

gusher *n:* an oil well that has come in with such great pressure that the oil jets out of the well like a geyser. In reality, a gusher is a blowout and is extremely wasteful of reservoir fluids and drive energy. In the early days of the oil industry, gushers were common and many times were the only indication that a large reservoir of oil and gas had been struck. See *blowout.*

H

hoist *n:* an arrangement of pulleys and wire rope or chain used for lifting heavy objects; a winch or similar device; the drawworks. See *drawworks.*

hoisting drum *n:* the large, flanged spool in the drawworks on which the hoisting cable is wound. See *drawworks.*

hook *n:* a large, hook-shaped device from which the swivel is suspended. It is designed to carry maximum loads ranging from 100 to 650 tons and turns on bearings in its supporting housing. A strong spring within the assembly cushions the weight of a stand (90 feet) of drill pipe, thus permitting the pipe to be made up and broken out with less damage to the tool joint threads. Smaller hooks without the spring are used for handling tubing and sucker rods. See *stand* and *swivel.*

hopper *n:* a large funnel- or cone-shaped device into which dry components (as powdered clay or cement) can be poured in order to uniformly mix the components with water (or other liquids). The liquid is injected through a nozzle at the bottom of the hopper. The resulting mixture of dry material and liquid may be drilling mud to be used as the circulating fluid in a rotary drilling operation or may be cement slurry used to bond casing to the borehole.

hydraulic fracturing *n:* an operation in which a specially blended liquid is pumped down a well and into a formation under pressure high enough to cause the formation to crack open. The resulting cracks or fractures serve as passages through which oil can flow into the wellbore. See *formation fracturing.*

hydrocarbons *n:* organic compounds of hydrogen and carbon, whose densities, boiling points, and freezing points increase as their molecular weights increase. Although composed only of two elements, hydrocarbons exist in a variety of compounds, because of the strong affinity of the carbon atom for other atoms and for itself. The smallest molecules of hydrocarbons are gaseous; the largest are solids. Petroleum is a mixture of many different hydrocarbons.

hydromatic brake *n:* a device mounted on the end of the drawworks shaft of a drilling rig. The hydromatic brake (often simply called the hydromatic) serves as an auxiliary to the mechanical brake when pipe is lowered into the well. The braking effect in a hydromatic brake is achieved by means of a runner or impeller turning in a housing filled with water.

I

impermeable *adj:* preventing the passage of fluid. A formation may be porous yet impermeable if there is an absence of connecting passages between the voids within it. See *permeability.*

inland barge rig *n:* a drilling structure consisting of a barge upon which the drilling equipment is constructed. When moved from one location to another, the barge floats, but, when stationed on the drill site, the barge is submerged to rest on the bottom. Typically, inland barge rigs are used to drill wells in marshes, shallow inland bays, and in areas where the water covering the drill site is not too deep.

instrumentation *n:* a device or assembly of devices designed for one or more of the following functions: to measure operating variables (as pressure, temperature, rate of flow, speed of rotation, etc.); to indicate these phenomena with visible or audible signals; to record them; to control them within a predetermined range; and to stop operations if the control fails. Simple instrumentation might consist of an indicating pressure gauge only. In a completely automatic system, the desired range of pressure, temperature, and so on is predetermined and preset.

intermediate casing string *n:* the string of casing set in a well after the surface casing, but before the production casing, to keep the hole from caving and to seal off troublesome formations. The string is sometimes called protection casing.

J

jackup drilling rig *n:* an offshore drilling structure with tubular or derrick legs that support the deck and hull. When positioned over the drilling site, the bottoms of the legs rest on the seafloor. A jackup rig is towed or propelled to a location with its legs up. Once the legs are firmly positioned on the bottom, the deck and hull height are adjusted and leveled.

jet bit *n:* a drilling bit having replaceable nozzles through with the drilling fluid is directed in a high-velocity stream to the bottom of the hole to improve the efficiency of the bit. See *bit*.

jet gun *n:* an assembly, including a carrier and shaped charges, that is used in jet perforating.

jet-perforate *v:* to create a hole through the casing with a shaped charge of high explosives instead of a gun that fires projectiles. The loaded charges are lowered into the hole to the desired depth. Once detonated, the charges emit short, penetrating jets of high-velocity gases that cut holes in the casing and cement and some distance into the formation. Formation fluids then flow into the wellbore through these perforations. See *bullet perforator* and *gun-perforate*.

joint *n:* a single length (about 30 feet) of drill pipe or of drill collar, casing, or tubing, that has threaded connections at both ends. Several joints screwed together constitute a stand of pipe. See *stand, single, double, thribble,* and *fourble*.

junk *n:* metal debris lost in a hole. Junk may be a lost bit, pieces of a bit, milled pieces of pipe, wrenches, or any relatively small object that impedes drilling and must be fished out of the hole. *v:* to abandon (as a nonproductive well).

K

kelly *n:* the heavy steel member, four- or six-sided, suspended from the swivel through the rotary table and connected to the topmost joint of drill pipe to turn the drill stem as the rotary table turns. It has a bored passageway that permits fluid to be circulated into the drill stem and up the annulus, or vice versa. See *drill stem, rotary table,* and *swivel*.

kelly bushing *n:* a special device that, when fitted into the master bushing, transmits torque to the kelly and simultaneously permits vertical movement of the kelly to make hole. It may be shaped to fit the rotary opening or have pins for transmitting torque. Also called the drive bushing. See *kelly* and *master bushing*.

kelly spinner *n:* a pneumatically operated device mounted on top of the kelly that, when actuated, causes the kelly to turn or spin. It is useful when the kelly or a joint of pipe attached to it must be spun up; that is, rotated rapidly in order to make it up.

kick *n:* an entry of water, gas, oil, or other formation fluid into the wellbore. It occurs because the pressure exerted by the column of drilling fluid is not great enough to overcome the pressure exerted by the fluids in the formation drilled. If prompt action is not taken to control the kick or kill the well, a blowout will occur. See *blowout*.

L

LACT unit *n:* an automated system for measuring and transferring oil from a lease gathering system into a pipeline. See *lease automatic custody transfer*.

latch on *v:* to attach elevators to a section of pipe to pull it out of or run it into the hole.

lead tongs *n:* (pronounced "leed") the pipe tongs suspended in the derrick or mast and operated by a wireline connected to the breakout cathead. Also called breakout tongs.

lease *n:* 1. a legal document executed between a landowner, as lessor, and a company or individual, as lessee, that grants the right to exploit the premises for minerals or other products. 2. the area where production wells, stock tanks, separators, LACT units, and other production equipment are located. See *LACT unit* and *lease automatic custody transfer*.

lease automatic custody transfer *n:* the measurement and transfer of oil from the producer's tanks to the connected pipeline on an automatic basis without a representative of either having to be present. See *LACT unit*.

location *n:* the place where a well is drilled.

log *n:* a systematic recording of data, as from the driller's log, mud log, electrical well log, or radioactivity log. Many different logs are run in wells being produced or drilled to obtain various characteristics of downhole formations.

M

magnetic brake *n:* also called an electrodynamic brake. See *electrodynamic brake*.

make a connection *v:* to attach a joint of drill pipe onto the drill stem suspended in the wellbore to permit deepening of the wellbore.

make a trip *v:* to hoist the drill stem out of the wellbore to perform one of a number of operations such as changing bits, taking a core, and so forth, and then to return the drill stem to the wellbore.

make hole *v:* to deepen the hole made by the bit; to drill ahead.

make up *v:* 1. to assemble and join parts to form a complete unit (as to make up a string of casing). 2. to screw together two threaded pieces. 3. to mix or prepare (as to make up a tank of mud). 4. to compensate for (as to make up for lost time).

make up a joint *v:* to screw a length of pipe into another length of pipe.

makeup cathead *n:* a device attached to the shaft of the drawworks that is used as a power source for screwing together joints of pipe; usually located on the driller's side of the drawworks. See *cathead*.

mast *n:* a portable derrick capable of being erected as a unit, as distinguished from a standard derrick that cannot be raised to a working position as a unit. For transporting by land, the mast can be divided into two or more sections to avoid excessive length extending from truck beds on the highway. Compare *derrick*.

master bushing *n:* a device that fits into the rotary table. It accommodates the slips and drives the kelly bushing so that the rotating motion of the rotary table can be transmitted to the kelly. Also called rotary bushing. See *slips* and *kelly bushing.*

mechanical rig *n:* a drilling rig in which the source of power is one or more internal-combustion engines and in which the power is distributed to rig components through mechanical devices (as chains, sprockets, clutches, and shafts). It is also called a power rig.

mill *n:* a downhole tool with rough, sharp, extremely hard cutting surfaces for removing metal by grinding or cutting. Mills are run on drill pipe or tubing to grind up debris in the hole, remove stuck portions of drill stem or sections of casing for sidetracking, and ream out tight spots in the casing. They are also called junk mills, reaming mills, and so forth, depending on what use they have. *v:* to use a mill to cut or grind metal objects that must be removed from a well.

mix mud *v:* to prepare drilling fluids from a mixture of water or other fluids and one or more of the various dry mud-making materials (as clay, weighting materials, chemicals, etc.).

monkeyboard *n:* the derrickman's working platform. As pipe or tubing is run into or out of the hole, the derrickman must handle the top end of the pipe, which may be as high as 90 feet in the derrick or mast. The monkeyboard provides a small platform to raise him to the proper height to be able to handle the top of the pipe. See *double board, fourble board,* and *thribble board.*

morning tour *n:* (pronounced "tower") also called daylight tour. See *daylight tour.*

motorman *n:* the crew member on a rotary drilling rig responsible for the care and operation of drilling engines.

mousehole *n:* an opening through the rig floor, usually lined with pipe, into which a length of drill pipe is placed temporarily for later connection to the drill string.

mousehole connection *n:* the procedure of adding a length of drill pipe or tubing to the active string in which the length to be added is placed in the mousehole, made up to the kelly, then pulled out of the mousehole, and subsequently made up into the string.

mud *n:* the liquid circulated through the wellbore during rotary drilling operations. In addition to its function of bringing cuttings to the surface, drilling mud cools and lubricates the bit and drill stem, protects against blowouts by holding back subsurface pressures, and deposits a mud cake on the wall of the borehole to prevent loss of fluids to the formation. Although it originally was a suspension of earth solids (especially clays) in water, the mud used in modern drilling operations is a more complex, three-phase mixture of liquids, reactive solids, and inert solids. The liquid phase may be fresh water, diesel oil, or crude oil and may contain one or more conditioners. See *drilling fluid.*

mud analysis *n:* examination and testing of the drilling mud to determine its physical and chemical properties.

mud cake *n:* the sheath of mud solids that forms on the wall of the hole when the liquid from the mud filters into the formation; also called wall cake or filter cake.

mud circulation *n:* the act of pumping mud downward to the bit and back up to the surface by normal circulation or reverse circulation. See *normal circulation* and *reverse circulation.*

mud conditioning *n:* the treatment and control of drilling mud to ensure that is has the correct properties. Condition-ing may include the use of additives, the removal of sand or other solids, the removal of gas, the addition of water, and other measures to prepare the mud for conditions encountered in a specific well.

mud engineer *n:* a person whose duty is to test and maintain the properties of the drilling mud that are specified by the operator.

mud gun *n:* a pipe that shoots a jet of drilling mud under high pressure into the mud pit to mix additives with the mud.

mud logging *n:* the recording of information derived from examination and analysis of formation cuttings made by the bit and mud circulated out of the hole. A portion of the mud is diverted through a gas-detecting device. Cuttings brought up by the mud are examined under ultraviolet light to detect the presence of oil or gas. Mud logging is often carried out in a portable laboratory set up at the well.

mud man *n:* also called a mud engineer. See *mud engineer.*

mud pit *n:* a series of open tanks, usually made of steel plates, through which the drilling mud is cycled to allow sand and sediments to settle out. Additives are mixed with the mud in the pit, and the fluid is temporarily stored there before being pumped back into the well. Modern rotary drilling rigs are generally provided with three or more pits, usually fabricated steel tanks fitted with built-in piping, valves, and mud agitators. Mud pits are also called shaker pits, settling pits, and suction pits, depending of their main purpose. See *shaker pit, settling pit,* and *suction pit.*

mud pump *n:* a large, reciprocating pump used to circulate the mud on a drilling rig. A typical mud pump is a single- or double-acting, two- or three-cylinder piston pump whose pistons travel in replaceable liners and are driven by a crankshaft actuated by an engine or motor. Also called a slush pump.

mud-return line *n:* a trough or pipe placed between the surface connections at the wellbore and the shale shaker, through which drilling mud flows upon its return to the surface from the hole.

mud screen *n:* also called a shale shaker. See *shale shaker.*

N

natural gas *n:* a highly compressible, highly expansible mixture of hydrocarbons having a low specific gravity and occurring naturally in a gaseous form. Besides hydrocarbon gases, natural gas may contain appreciable quantities of nitrogen, helium, carbon dioxide, and contaminants (as hydrogen sulfide and water vapor). Although gaseous at normal temperatures and pressures, certain of the gases comprising the mixture that is natural gas are variable in form and may be found either as gases or as liquids under suitable conditions of temperature and pressure.

needle valve *n:* a globe valve that incorporates a needle-point disk to produce extremely fine regulation of flow.

nipple *n:* a tubular pipe fitting threaded on both ends and less than 12 inches long.

nipple up *v:* in drilling, to assemble the blowout-preventer stack on the wellhead at the surface.

normal circulation *n:* the smooth, uninterrupted circulation of drilling fluid down the drill stem, out the bit, up the annular space between the pipe and the hole, and back to the surface. See *mud circulation* and *reverse circulation.*

O

offshore drilling *n:* drilling for oil in an ocean, gulf, or sea, usually on the continental shelf. A drilling unit for offshore operations may be a mobile floating vessel with a ship or barge hull, a semisubmersible or submersible base, a self-propelled or towed structure with jacking legs (jackup drilling rig), or a permanent structure used as a production platform when drilling is completed. In general, wildcat wells are drilled from mobile floating vessels (as semisubmerisble rigs and drill ships) or from jackups, while development wells are drilled from platforms. See *drill ship, jackup drilling rig, platform, semisubmersible drilling rig,* and *wildcat.*

oil field *n:* the surface area overlying an oil reservoir or reservoirs. Commonly, the term includes not only the surface area, but may include the reservoir, the wells, and production equipment as well.

oil sand *n:* 1. a sandstone that yields oil. 2. (by extension) any reservoir that yields oil, whether or not it is sandstone.

oil zone *n:* a formation or horizon of a well from which oil may be produced. The oil zone is usually immediately under the gas zone and on top of the water zone if all three fluids are present and segregated.

open *adj:* 1. of a wellbore, having no casing. 2. of a hole, having no drill pipe or tubing suspended in it.

open hole *n:* 1. any wellbore in which casing has not been set. 2. open or cased hole in which no drill pipe or tubing is suspended.

operator *n:* the person or company, either proprietor or lessee, actually operating an oilwell or lease. Compare *unit operator.*

overshot *n:* a fishing tool that is attached to tubing or drill pipe and lowered over the outside wall of pipe lost or stuck in the wellbore. A friction device in the overshot, usually either a basket or a spiral grapple, firmly grips the pipe, allowing the lost fish to be pulled from the hole.

P

P&A *abbr:* plug and abandon.

pay sand *n:* the producing formation, often one that is not even sandstone. It is also called pay, pay zone, and producing zone.

perforate *v:* to pierce the casing wall and cement to provide holes through which formation fluids may enter or to provide holes in the casing so that materials may be introduced into the annulus between the casing and the wall of the borehole. Perforating is accomplished by lowering into the well a perforating gun, or perforator, that fires electrically detonated bullets or shaped charges from the surface. See *perforating gun.*

perforating gun *n:* a device fitted with shaped charges or bullets that is lowered to the desired depth in a well and fired to create penetrating holes in casing, cementing, and formation. See *gun-perforate.*

permeability *n:* 1. a measure of the ability of fluids to flow through a porous rock. 2. fluid conductivity of a porous medium. 3. the ability of a fluid to flow within the interconnected pore network of a porous medium. See *absolute permeability, effective permeability,* and *relative permeability.*

petroleum *n:* oil or gas obtained from the rocks of the earth. See *hydrocarbons.*

pin *n:* the male section of the tool joint. See *tool joint.*

pipe ram *n:* a sealing component for a blowout preventer that closes the annular space between the pipe and the blowout preventer or wellhead. See *annular space* and *blowout preventer.*

platform *n:* an immobile, offshore structure constructed on pilings from which wells are drilled, produced, or both.

plug and abandon *v:* to place a cement plug into a dry hole and abandon it.

pore *n:* an opening or space within a rock or mass of rocks, usually small and often filled with some fluid (as water, oil, gas, or all three). Compare *vug.*

porosity *n:* the condition of something that contains pores (as a rock formation). See *pore.*

positive choke *n:* a choke in which the orifice size must be changed to change the rate of flow through the choke. See *choke* and *orifice.*

pressure *n:* the force that a fluid (liquid or gas) exerts when it is in some way confined within a vessel, pipe, hole in the ground, and so forth, such as that exerted against the inner wall of a tank or that exerted on the bottom of the wellbore by drilling mud. Pressure is often expressed in terms of force per unit of area, as pounds per square inch (psi).

pressure gauge *n:* an instrument for measuring fluid pressure that usually registers the difference between atmospheric pressure and the pressure of the fluid by indicating the effect of such pressures on a measuring element (as a column of liquid, a weighted piston, a diaphragm, or other pressure-sensitive device).

pressure gradient *n:* a scale of pressure differences in which there is a uniform variation of pressure from point to point. For example, the pressure gradient of a column of water is about 0.433 psi/ft of vertical elevation (9.794 kPa/m). The normal pressure gradient in a well is equivalent to the pressure exerted at any given depth by a column of 10 percent salt water extending from that depth to the surface (i.e., 0.465 psi/ft or 10.518 kPa/m).

pressure relief valve *n:* a valve that opens at a preset pressure to relieve excessive pressures within a vessel or line; also called a relief valve, safety valve, or pop valve.

preventer *n:* shortened form of blowout preventer. See *blowout preventer.*

primary cementing *n:* the cementing operation that takes place immediately after the casing has been run into the hole; used to provide a protective sheath around the casing, to segregate the producing formation, and to prevent the migration of undesirable fluids. See *secondary cementing* and *squeeze cementing.*

prime mover *n:* an internal-combustion engine that is the source of power for a drilling rig in oilwell drilling.

production *n:* 1. the phase of the petroleum industry that deals with bringing the well fluids to the surface and separating them and with storing, gauging, and otherwise preparing the product for the pipeline. 2. the amount of oil or gas produced in a given period.

proppant *n:* also called propping agent. See *propping agent.*

propping agent *n:* a granular substance (as sand grains, aluminum pellets, or other material) carried in suspension by the fracturing fluid that serves to keep the cracks open when the fracturing fluid is withdrawn after a fracture treatment.

psi *abbr:* pounds per square inch. See *pressure.*

pump *n:* a device that increases the pressure on a fluid or raises it to a higher level. Various types of pumps include the reciprocating pump, centrifugal pump, rotary pump, jet pump, sucker rod pump, hydraulic pump, mud pump, submersible pump, and bottomhole pump.

pump pressure *n:* fluid pressure arising from the action of the pump.

R

radioactivity well logging *n:* the recording of the natural or induced radioactive characteristics of subsurface formations. A radioactivity log, also known as a radiation log, normally consists of two recorded curves: a gamma ray curve and a neutron curve. Both indicate the types of rocks in the formation and the types of fluids contained in the rocks. The two logs may be run simultaneously in conjunction with a collar locator in a cased or uncased hole.

ram *n:* the closing and sealing component on a blowout preventer. One of three types—blind, pipe, or shear—may be installed in several preventers mounted in a stack on top of the wellbore. Blind rams, when closed, form a seal on a hole that has no drill pipe in it; pipe rams, when closed, seal around the pipe; shear rams cut through drill pipe and then form a seal. See *blind ram, pipe ram,* and *shear ram.*

ram blowout preventer *n:* a blowout preventer that uses rams to seal off pressure on a hole that is with or without pipe. It is also called a ram preventer. See *blowout preventer* and *ram.*

rathole *n:* 1. a hole in the rig floor 30 to 35 feet deep, lined with casing that projects above the floor, into which the kelly and swivel are placed when hoisting operations are in progress. 2. a hole of a diameter smaller than the main hole that is drilled in the bottom of the main hole. *v:* to reduce the size of the wellbore and drill ahead.

reeve *v:* to pass (as the end of a rope) through a hole or opening in a block or similar device.

reeve the line *v:* to string a wire-rope drilling line through the sheaves of the traveling and crown blocks to the hoisting drum.

relative permeability *n:* a measure of the ability of two or more fluids (as water, gas, and oil) to flow through a rock formation when the formation is totally filled with several fluids. The permeability measure of a rock filled with two or more fluids is different from the permeability measure of the same rock filled with only a single fluid. Compare *absolute permeability.*

reserve pit *n:* 1. (obsolete) a mud pit in which a supply of drilling fluid was stored. 2. a waste pit, usually an excavated, earthen-walled pit. It may be lined with plastic to prevent contamination of the soil.

reservoir *n:* a subsurface, porous, permeable rock body in which oil and/or gas is stored. Most reservoir rocks are limestones, dolomites, sandstones, or a combination of these. The three basic types of hydrocarbon reservoirs are oil, gas, and condensate. An oil reservoir generally contains three fluids—gas, oil, and water—with oil the dominant product. In the typical oil reservoir, these fluids occur in different phases because of the variance in their gravities. Gas, the lightest, occupies the upper part of the reservoir rocks; water, the lower part; and oil, the intermediate section. In addition to occurring as a cap or in solution, gas may accumulate independently of the oil; if so, the reservoir is called a gas reservoir. Associated with the gas, in most instances, are salt water and some oil. In a condensate reservoir, the hydrocarbons may exist as a gas, but, when brought to the surface, some of the heavier ones condense to a liquid or condensate.

reservoir pressure *n:* the pressure in a reservoir under normal conditions.

reverse circulation *n:* the return of drilling fluid through the drill stem. The normal course of drilling fluid circulation is downward through the drill stem and upward through the annular space surrounding the drill stem. For special problems, normal circulation is sometimes reversed, and the fluid returns to the surface through the drill stem, or tubing, after being pumped down the annulus.

rig *n:* the derrick or mast, drawworks, and attendant surface equipment of a drilling unit.

rig down *v:* to dismantle the drilling rig and auxiliary equipment following the completion of drilling operations; also called tear down.

rig up *v:* to prepare the drilling rig for making hole; to install tools and machinery before drilling is started.

roller cone bit *n:* a drilling bit made of two, three, or four cones, or cutters, that are mounted on extremely rugged bearings. Also called rock bits. The surface of each cone is made up of rows of steel teeth or rows of tungsten carbide inserts. See *bit.*

rotary bushing *n:* also called master bushing. See *master bushing.*

rotary drilling *n:* a drilling method in which a hole is drilled by a rotating bit to which downward force is applied. The bit is fastened to and rotated by the drill stem, which also provides a passageway through which the drilling fluid is circulated. Additional joints of drill pipe are added as drilling progresses.

rotary helper *n:* a worker on a drilling rig, subordinate to the driller; sometimes called a roughneck, floorman, or rig crewman.

rotary hose *n:* a reinforced, flexible tube on a rotary drilling rig that conducts the drilling fluid from the mud pump and standpipe to the swivel and kelly; also called the mud hose or the kelly hose. See *kelly, mud pump, standpipe,* and *swivel.*

rotary table *n:* the principal component of a rotary, or rotary machine, used to turn the drill stem and support the drilling assembly. It has a beveled gear arrangement to create the rotational motion and an opening into which bushings are fitted to drive and support the drilling assembly.

roughneck *n:* also called a rotary helper. See *rotary helper.*

round trip *n:* the action of pulling out and subsequently running back into the hole a string of drill pipe or tubing. It is also called tripping.

roustabout *n:* 1. a worker on an offshore rig who handles the equipment and supplies that are sent to the rig from the shore base. The head roustabout is very often the crane operator. 2. a worker who assists the foreman in the general work around a producing oil well, usually on the property of the oil company. 3. a helper on a well-servicing unit.

run in *v:* to go into the hole with tubing, drill pipe, and so forth.

S

samples *n pl:* 1. the well cuttings obtained at designated footage intervals during drilling. From an examination of these cuttings, the geologist determines the type of rock and formations being drilled and estimates oil and gas content. 2. small quantities of well fluids obtained for analysis.

sand *n:* 1. an abrasive material composed of small quartz grains formed from the disintegration of preexisting rocks. Sand consists of particles less than 2 millimetres and greater than 1/16 of a millimetre in diameter. 2. sandstone.

scratcher *n:* a device fastened to the outside of casing that removes the mud cake from the wall of the hole to condition the hole for cementing. By rotating or moving the casing string up and down as it is being run into the hole, the scratcher, formed of stiff wire, removes the cake so that the cement can bond solidly to the formation.

secondary cementing *n:* any cementing operation after the primary cementing operation. Secondary cementing includes a plug-back job, in which a plug of cement is positioned at a specific point in the well and allowed to set. Wells are plugged to shut off bottom water or to reduce the depth of the well for other reasons. See *primary cementing* and *squeeze cementing.*

seismograph *n:* a device that detects reflections of vibrations in the earth, used in prospecting for probable oil-bearing structures. Vibrations are created by discharging explosives in shallow boreholes, by striking the surface with a heavy blow, or by generating low-frequency sound waves. The type and velocity of the vibrations as recorded by the seismograph indicate the general characteristics of the section of earth through which the vibrations pass.

semisubmersible drilling rig *n:* a floating, offshore drilling structure that has hulls submerged in the water but not resting on the seafloor. Living quarters, storage space, and so forth are assembled on the deck. Semisubmersible rigs are either self-propelled or towed to a drilling site and either anchored or dynamically positioned over the site or both. Semisubmersibles are more stable than drill ships and are used extensively to drill wildcat wells in rough waters such as the North Sea. See *dynamic positioning.*

set casing *v:* to run and cement casing at a certain depth in the wellbore. Sometimes, the term "set pipe" is used when referring to setting casing.

settling pit *n:* the mud pit into which mud flows and in which heavy solids are allowed to settle out. Often auxiliary equipment (as desanders) must be installed to speed this process.

shaker *n:* shortened form of shale shaker. See *shale shaker.*

shaker pit *n:* the mud pit adjacent to the shale shaker, usually the first pit into which the mud flows after returning from the hole.

shale *n:* a fine-grained sedimentary rock composed of consolidated silt and clay or mud. Shale is the most frequently occurring sedimentary rock.

shale shaker *n:* a series of trays with sieves that vibrate to remove cuttings from the circulating fluid in rotary drilling operations. The size of the openings in the sieve is carefully selected to match the size of the solids in the drilling fluid and the anticipated size of cuttings. Also called a shaker.

shaped charge *n:* a relatively small container of high explosive that is loaded into a perforating gun. Upon detonation, the charge releases a small, high-velocity stream of particles (a jet) that penetrates the casing, cement, and formation. See *gun-perforate.*

shear ram *n:* the components in a blowout preventer that cut, or shear, through drill pipe and form a seal against well pressure. Shear rams are used in mobile offshore drilling operations to provide a quick method of moving the rig away from the hole when there is no time to trip the drill stem out of the hole.

sheave *n:* (pronounced "shiv") a grooved pulley.

show *n:* the appearance of oil or gas in cuttings, samples, cores, and so forth of drilling mud.

shut down *v:* to stop work temporarily or to stop a machine or operation.

shut-in bottomhole pressure *n:* the pressure at the bottom of a well when the surface valves on the well are completely closed. The pressure is caused by fluids that exist in the formation at the bottom of the well.

sidetrack *v:* to drill around broken drill pipe or casing that has become lodged permanently in the hole, using a whipstock, turbodrill, or other mud motor. See *directional drilling, turbodrill,* and *whipstock.*

sidewall coring *n:* a coring technique in which core samples are obtained from a zone that has already been drilled. A hollow bullet is fired into the formation wall to capture the core and then retrieved on a flexible steel cable. Core samples of this type usually range from ¾ to 1³/16 inches in diameter and from ¾ to 1 inch in length. This method is especially useful in soft rock areas.

single *n:* a joint of drill pipe. Compare *double, thribble,* and *fourble.*

slips *n pl:* wedge-shaped pieces of metal with teeth or other gripping elements that are used to prevent pipe from slipping down into the hole or to hold pipe in place. Rotary slips fit around the drill pipe and wedge against the master bushing to support the pipe. Power slips are pneumatically or hydraulically actuated devices that allow the crew to dispense with the manual handling of slips when making a connection. Packers and other downhole equipment are secured in position by slips that engage the pipe by action directed at the surface.

slurry *n:* a plastic mixture of cement and water that is pumped into a well to harden; there it supports the casing and provides a seal in the wellbore to prevent migration of underground fluids.

sonic logging *n:* the recording of the time required for a sound wave to travel a specific distance through a formation. Difference in observed travel times is largely caused by variations in porosities of the medium, an important determination. The sonic log, which may be run simultaneously with a spontaneous potential log or a gamma ray log, is useful for correlation and often is used in conjunction with other logging services for substantiation of porosities. It is run in an uncased hole.

spear *n:* a fishing tool used to retrieve pipe lost in a well. The spear is lowered down the hole and into the lost pipe, and, when weight, torque, or both are applied to the string to which the spear is attached, the slips in the spear expand and tightly grip the inside of the wall of the lost pipe. Then the string, spear, and lost pipe are pulled to the surface.

specific gravity *n:* the ratio of the weight of a given volume of a substance at a given temperature to the weight of an equal volume of a standard substance at the same temperature. For example, if 1 cubic inch of water at 39°F weighs 1 unit and 1 cubic inch of another solid or liquid at 39°F weighs 0.95 unit, then the specific gravity of the substance is 0.95. In determining the specific gravity of gases, the comparison is made with the standard of air or hydrogen. See *gravity.*

spinning cathead *n:* a spooling attachment on the makeup cathead to permit use of a spinning chain to spin up or make up drill pipe. See *spinning chain.*

spinning chain *n:* a Y-shaped chain used to spin up (tighten) one joint of drill pipe into another. In use, one end of the chain is attached to the tongs, another end to the spinning cathead, and the third end is free. The free end is wrapped around the tool joint, and the cathead pulls the chain off the joint, causing the joint to spin (turn) rapidly and tighten up. After the chain is pulled off the joint, the tongs are secured in the same spot, and continued pull on the chain (and thus on the tongs) by the cathead makes up the joint to final tightness.

spud *v:* to move the drill stem up and down in the hole over a short distance without rotation. Careless execution of this operation creates pressure surges that can cause a formation to break down, which results in lost circulation. See *spud in.*

spud in *v:* to begin drilling; to start the hole.

squeeze cementing *n:* the forcing of cement slurry by pressure to specified points in a well to cause seals at the points of squeeze. It is a secondary cementing method that is used to isolate a producing formation, seal off water, repair casing leaks, and so forth. See *cementing.*

stab *v:* to guide the end of a pipe into a coupling or tool joint when making up a connection. See *coupling* and *tool joint.*

stabbing board *n:* a temporary platform erected in the derrick or mast some 20 to 40 feet above the derrick floor. The derrickman or another crew member works on the board while casing is being run in a well. The board may be wooden or fabricated of steel girders floored with antiskid material and powered electrically to raise or lower it to the desired level. A stabbing board serves the same purpose as a monkeyboard but is temporary instead of permanent.

stake a well *v:* to locate precisely on the surface of the ground the point at which a well is to be drilled. After exploration techniques have revealed the possibility of the existence of a subsurface, hydrocarbon-bearing formation, a certified and registered land surveyor drives a stake into the ground to mark the spot where the well is to be drilled.

stand *n:* the connected joints of pipe racked in the derrick or mast when making a trip. On a rig, the usual stand is 90 feet long (three lengths of pipe screwed together) or a thribble. Compare *double* and *fourble.*

standpipe *n:* a vertical pipe rising along the side of the derrick or mast, which joins the discharge line leading from the mud pump to the rotary hose and through which mud is pumped going into the hole. See *mud pump* and *rotary hose.*

stimulation *n:* any process undertaken to enlarge old channels or create new ones in the producing formation of a well (e.g., acidizing or formation fracturing). See *acidize.*

stratification *n:* the natural layering or lamination characteristic of sediments and sedimentary rocks.

stratigraphic trap *n:* a petroleum trap that occurs when the top of the reservoir bed is terminated by other beds or by a change of porosity or permeability within the reservoir itself. Compare *structural trap.* See *trap.*

string *n:* the entire length of casing, tubing, or drill pipe run into a hole; the casing string. Compare *drill string* and *drill stem.*

string up *v:* to thread the drilling line through the sheaves of the crown block and traveling block. One end of the line is

secured to the hoisting drum and the other to the derrick substructure. See *sheave.*

structural trap *n:* a petroleum trap that is formed because of deformation (as folding or faulting) of the rock layer that contains petroleum. Compare *stratigraphic trap.* See *trap.*

stuck pipe *n:* drill pipe, drill collars, casing, or tubing that has inadvertently become immobile in the hole. It may occur when drilling is in progress, when casing is being run in the hole, or when the drill pipe is being hoisted.

sub *n:* a short, threaded piece of pipe used to adapt parts of the drilling string that cannot otherwise be screwed together because of differences in thread size or design. A sub may also perform a special function. Lifting subs are used with drill collars to provide a shoulder to fit the drill pipe elevators. A kelly saver sub is placed between the drill pipe and kelly to prevent excessive thread wear of the kelly and drill pipe threads. A bent sub is used when drilling a directional hole. Sub is a short expression for substitute.

submersible drilling rig *n:* an offshore drilling structure with several compartments that are flooded to cause the structure to submerge and rest on the seafloor. Most submersible rigs are used only in shallow waters.

substructure *n:* the foundation on which the derrick or mast and usually the drawworks sit; contains space for storage and well control equipment.

suction pit *n:* the mud pit from which mud is picked up by the suction of the mud pumps; also called a sump pit and mud suction pit.

surface casing *n:* also called surface pipe. See *surface pipe.*

surface pipe *n:* the first string of casing (after the conductor pipe) that is set in a well, varying in length from a few hundred to several thousand feet. Some states require a minimum length to protect freshwater sands. Compare *conductor pipe.*

swivel *n:* a rotary tool that is hung from the rotary hook and traveling block to suspend and permit free rotation of the drill stem. It also provides a connection for the rotary hose and a passageway for the flow of drilling fluid into the drill stem.

syncline *n:* a downwarped, trough-shaped configuration of folded, stratified rocks. Compare *anticline.*

T

TD *abbr:* total depth.

thread protector *n:* a device that is screwed onto or into pipe threads to protect the threads from damage when the pipe is not in use. Protectors may be metal or plastic.

thribble *n:* a stand of pipe made up of three joints and handled as a unit. See *stand.* Compare *single, double,* and *fourble.*

thribble board *n:* the name used for the working platform of the derrickman, or monkeyboard, when it is located at a height in the derrick equal to three lengths of pipe joined together. Compare *double board* and *fourble board.* See *monkeyboard.*

throw the chain *n:* to flip the spinning chain up from a tool joint box so that the chain wraps around the tool joint pin after it is stabbed into the box. The stand or joint of drill pipe is turned or spun by a pull on the spinning chain from the cathead on the drawworks.

tight formation *n:* a petroleum- or water-bearing formation of relatively low porosity and permeability. See *porosity* and *permeability.*

tight hole *n:* a well about which information is restricted and passed only to those authorized for security or competitive reasons.

tongs *n pl:* the large wrenches used for turning when making up or breaking out drill pipe, casing, tubing, or other pipe; variously called casing tongs, rotary tongs, and so forth according to the specific use. Power tongs are pneumatically or hydraulically operated tools that serve to spin the pipe up tight and, in some instances, to apply the final makeup torque. See also *chain tongs.*

tool joint *n:* a heavy coupling element for drill pipe made of special alloy steel. Tool joints have coarse, tapered threads and seating shoulders designed to sustain the weight of the drill stem, withstand the strain of frequent coupling and uncoupling, and provide a leakproof seal. The male section of the joint, or the pin, is attached to one end of a length of drill pipe, and the female section, or box, is attached to the other end. The tool joint may be welded to the end of the pipe or screwed on or both. A hard metal facing is often applied in a band around the outside of the tool joint to enable it to resist abrasion from the walls of the borehole.

tool pusher *n:* an employee of a drilling contractor who is in charge of the entire drilling crew and the drilling rig. Also called a drilling rig foreman, manager, supervisor, or rig superintendent. See *drilling foreman.*

torque *n:* the turning force that is applied to a shaft or other rotary mechanism to cause it to rotate or tend to do so. Torque is measured in foot-pounds, joules, meter-kilograms, and so forth.

torque converter *n:* a connecting device between a prime mover and the machine actuated by it. The elements that pump the fluid in the torque converter automatically increase the output torque of the engine to which the torque is applied, with an increase of load on the output shaft. Torque converters are used extensively on mechanical rigs that have a compound. See *mechanical rig.*

total depth *n:* the maximum depth reached in a well.

tour *n:* (pronounced "tower") an 8-hour shift worked by a drilling crew or other oil field workers. Sometimes 12-hour tours are used, especially on offshore rigs. The most common divisions of tours are daylight, evening, and graveyard, if 8-hour tours are employed.

transmission *n:* the gear or chain arrangement by which power is transmitted from the prime mover to the drawworks, mud pump, or rotary table of a drilling rig. See *prime mover.*

trap *n:* layers of buried rock strata that are arranged so that petroleum accumulates in them.

traveling block *n:* an arrangement of pulleys, or sheaves, through which drilling cable is reeved and that moves up and down in the derrick or mast. See *block, crown block,* and *sheave.*

tricone bit *n:* a type of bit in which three cone-shaped cutting devices are mounted in such a way that they intermesh and rotate together as the bit drills. The bit body may be fitted with nozzles, or jets, through which the drilling fluid is discharged. A one-eyed bit is used in soft formations to drill a deviated hole. See *directional drilling* and *bit.*

trip *n:* the operation of hoisting the drill stem from and returning it to the wellbore. See *make a trip.*

turbodrill *n:* a drilling tool that rotates a bit attached to it by the action of the drilling mud on the turbine blades built into the tool. When a turbodrill is used, rotary motion is imparted only at the bit; therefore, it is unnecessary to rotate the drill stem. Although straight holes can be drilled with the tool, it is used most often in directional drilling.

U

unit operator *n:* the oil company in charge of development and producing in an oil field in which several companies have joined together to produce the field.

V

valve *n:* a device used to control the rate of flow in a line, to open or shut off a line completely, or to serve as an automatic or semiautomatic safety device. Those with extensive usage include the gate valve, plug valve, globe valve, needle valve, check valve, and pressure relief valve. See *check valve, needle valve,* and *pressure relief valve.*

V-belt *n:* a belt with a trapezoidal cross section that is made to run in sheaves, or pulleys, with grooves of corresponding shape. See *belt.*

vug *n:* a cavity in a rock.

W

waiting on cement *adj:* pertaining to or during the time when drilling or completion operations are suspended so the cement in a well can harden sufficiently.

wall cake *n:* also called filter cake and mud cake. See *filter cake* and *mud cake.*

weevil *n:* shortened form of boll weevil. See *boll weevil.*

weight indicator *n:* an instrument near the driller's position on a drilling rig. It shows both the weight of the drill stem that is hanging from the hook (hook load) and the weight that is placed on the bit by the drill collars (weight on bit).

weighting material *n:* a material that has high specific gravity and is used to increase the density of drilling fluids or cement slurries.

wellbore *n:* a borehole; the hole drilled by the bit. A wellbore may have casing in it or may be open (i.e., uncased), or a portion of it may be cased and a portion of it may be open. Also called borehole or hole. See *cased* and *open.*

well completion *n:* the activities and methods necessary to prepare a well for the production of oil and gas; the method by which a flow line for hydrocarbons is established between the reservoir and the surface. The method of well completion used by the operator depends on the individual characteristics of the producing formation or formations. These techniques include open-hole completions, sand exclusion completions, tubingless completions, multiple completions, and miniaturized completions.

wellhead *n:* the equipment installed at the surface of the wellbore. A wellhead includes such equipment as the casinghead and tubing head. *adj:* pertaining to the wellhead (as wellhead pressure).

well logging *n:* the recording of information about subsurface geologic formations. Logging methods include records kept by the driller, mud and cutting analyses, core analysis, drill stem tests, and electric and radioactivity procedures. See *electric well log, mud logging, radioactivity well logging,* and *sonic logging.*

well stimulation *n:* any of several operations used to increase the production of a well. See *acidize* and *formation fracturing.*

whipstock *n:* a long, steel casing that uses an inclined plane to cause the bit to deflect from the original borehole at a slight angle. Whipstocks are sometimes used in controlled directional drilling, to straighten crooked boreholes, and to sidetrack to avoid unretrieved fish. See *directional drilling, fish,* and *sidetrack.*

wildcat *n:* 1. a well drilled in an area where no oil or gas production exists. With present-day exploration methods and equipment, about one wildcat out of every nine proves to be productive although not necessarily profitable. 2. (nautical) a geared sheave of a windlass used to pull anchor chain. *v:* to drill wildcat wells.

wireline *n:* a slender, rodlike or threadlike piece of metal, usually small in diameter, that is used for lowering special tools (such as logging sondes, perforating guns, and so forth) into the well. Compare *wire rope.*

wire rope *n:* a cable composed of steel wires twisted around a central core of hemp or other fiber to create a rope of great strength and considerable flexibility. Wire rope is used as drilling line (in rotary and cable-tool rigs), coring line, servicing line, winch line, and so on. It is often called cable or wireline; however, wireline is a single, slender metal rod, usually very flexible. Compare *wireline.*

WOC *abbr:* waiting on cement.

worm *n:* a new and inexperienced worker on a drilling rig.